I0760163

THE CONSTRUCTION

OF

LARGE INDUCTION COILS

A WORKSHOP HANDBOOK

BY

A. T. HARE, M.A.

LATE SCHOLAR OF WADHAM COLLEGE, OXFORD

WITH THIRTY-FIVE ILLUSTRATIONS

METHUEN & CO.

36 ESSEX STREET W.C.

LONDON

1900

D. VAN NOSTRAND COMPANY,

NEW YORK.

PREFACE

THE widespread interest which was excited early in 1896 by Professor Röntgen's discovery of some properties of the radiation from an exhausted tube has directed the attention of many students of physics to the construction of large induction coils. The writer, among the number, was drawn into this fascinating pursuit, and, doubtless like many more, was disappointed to find that the literature of the subject was very inadequate and difficult of access. Coil-making is an operation abounding in pitfalls for the unwary, and for obvious reasons those professionally engaged in it, who have themselves found means of overcoming the difficulties, are not anxious to make the result of their labours public. This may account for the fact that so little help is to be obtained from published handbooks. Practically the only useful information the writer found was in back numbers of the *English Mechanic*. What published books are to be had deal generally with the construction of all kinds of coils, medical and other, and have consequently but little space to devote to the special difficulties which arise in dealing with the larger sizes of sparking coils. There are many precautions which may be safely neglected in a coil designed to give a two- or three-inch spark, but which in the case of one which is to give a spark a foot long and upwards become absolutely essential to success.

It is proposed in this little handbook to give in considerable detail instructions which, if faithfully followed, will guide the amateur electrical engineer to the successful completion of a coil giving a spark thirteen or fourteen inches long, which will suffice for most experiments with Röntgen Rays, Wireless Telegraphy, and the study of high vacua. It is believed that the method described is practically that adopted by the best-known makers. At all events it is one which has stood the test of experience, and nothing is anywhere described in the book which has not been tried by the writer and found to be successful. There are some minor points in which he considers that something has been introduced, if not new, at least unusual and useful, especially to those who, besides making an induction coil for practical use, are desirous of studying the action of the coil itself; such are the methods of bringing out the middle of the secondary wire by a central partition, the winding of the primary in independent layers, and the adoption of removable breaks. These variations from established custom are, of course, not necessary, but they are very convenient, and add little or nothing to the trouble of construction, while they enable the user of the coil to obtain an insight into its working, which he cannot obtain where the coil is set to do one thing only, however well that one thing may be done. Moreover there is nothing in these instructions inconsistent with the ordinary modes of coil-making, for which they will equally serve. For points of theory and the history of the induction coil the reader is referred to electrical text-books, among which perhaps may be specially mentioned Fleming's "Alternate Current Transformer." This handbook deals exclusively with the practical details of construction. The writer has to thank Mr. A. A. Campbell Swinton for re ising the section on Electrolytic Breaks and Mr. W. R. Pidgeon for many valuable suggestions.

CONTENTS

LIST OF ILLUSTRATIONS

THE CONSTRUCTION

OF

LARGE INDUCTION COILS

I

THE ORDER OF CONSTRUCTION

IT is assumed that the reader is familiar with the elementary facts of electricity, and has such knowledge of the theory and construction of the induction coil as is to be found in all elementary text-books, viz., that it consists of an iron core round which are coiled two bobbins of wire, a primary coil consisting of a short length of thick wire, and a secondary coil consisting of a considerable length of thin wire, and that by sending through the primary an interrupted current of low voltage and great quantity, currents are induced in the secondary of less quantity but far higher voltage. Any reader who is not familiar with the elements of the subject is referred to the numerous excellent text-books on electricity, nearly all of which contain the information in question. It is further assumed that, being possessed of this knowledge, he is desirous of constructing an induction coil large enough for all ordinary purposes. This handbook is designed to smooth his way and enable him to overcome the difficulties which the construction undoubtedly presents.

We will first consider the order in which the various parts of the coil can be most advantageously made. For obvious reasons it is best to begin with the iron core, and then to wind the primary coil. These having been finished, it is convenient to construct a rough deal stand with two "Y" supports, so that the primary may be laid on them parallel with the baseboard and at the height at which it is intended finally to lie above the top of the finished stand. The commutator, break, and condenser may then be proceeded with, and fitted to the deal stand, with all terminals, etc. making a temporary connection with the ends of the primary wire. This will be found very handy for testing, which operation must be constantly performed during the subsequent processes. The ebonite insulation of the primary can next be fitted, and the sections of the secondary (which are best made separately) can be put on and tested in position, while it is still possible to take them off again if any prove defective. Lastly, the core with the primary coil and its insulation can be taken off the stand and the secondary fitted on and jointed up, and the ends and cover attached. The permanent stand can then be made, and the coil, break, commutator, etc., will be all ready to be screwed down upon it in their proper positions.

Adhering, then, to this scheme of construction, we commence with the core.

II

THE CORE

THE cutting up, straightening, and arranging a large number of iron wires being a troublesome task, it is fortunate that bundles of straightened wire are to be found ready made of many sizes on the market. They purport to be, and no doubt are, of the best quality of iron, but considering how much of the performance of a coil depends upon the rapid demagnetisability of the iron, it is a prudent precaution to anneal the iron oneself. This may be simply done by putting the bundle into a length of gas-pipe large enough to hold it easily, and screwing caps on the two ends, in which condition it can be allowed to "soak" in the heat of a large fire and cool slowly as the fire goes out. The wires will generally be found softer after this treatment. The exact gauge of the wire is not very important. No. 22 is a satisfactory size.

The next thing is to fix on the dimensions of the intended core, and to bring the bundle to the right shape and size. The reader will of course have decided on the size of the coil he is intending to construct. When special measurements are given in this handbook, it is assumed that a really powerful coil is being aimed at, which, when its capabilities are fully brought out, will give a spark over a foot long. The measurements are taken from one sparking easily between its terminals, which are set 13½ in. apart. If only a small variation from this size is desired, there is no doubt that a proportionate increase or diminution of all the linear dimensions will lead to the desired result, but it has been thought better to give exact measurements from an existing instrument, and to leave the reader to modify them if desired. It may be assumed, then, that the core is to be $1\frac{1}{16}$ in. in diameter and 18 in. long. Probably the bundle could if necessary be cut to length with a fine hack-saw, but bundles of wire of the above length are to be had ready cut and straightened, so that this troublesome work need not be incurred.

To make the core cylindrical and of the right diameter, a piece of sheet brass is chucked on a wooden face plate in the lathe, and a hole turned in it exactly $1\frac{1}{16}$ in. in diameter on one face, and countersunk on the other, so as to make an easy entrance for the core. After annealing, the wires should be made up roughly into a cylindrical bundle, and a sheet of paper of the same length and width as the core may be inserted so as to form a diametral plane in it. This is designed to check the formation of eddy currents. The bundle is then loosely bound up with wire. This wire binding is then gradually removed from the bundle, starting from one end, and the brass plate coaxed over the bundle by squeezing and pushing. Some of the outside wires will probably have to be taken off to render this possible, the bundles as bought being more than ample in quantity; but enough should be left to fill the hole in the plate completely and tightly. As soon as the brass plate has been pushed a couple of inches along the core, cotton tape is wound helically over the projecting end, the beginning of the tape being made fast by sewing, and each turn of the tape just overlapping the last. In this way the brass plate is gradually worked from one end to the other, the binding wire being at the same time removed on the one side and the serving of tape proceeded with on the other. The result will be a cylindrical bundle very true in form and size.

The whole is then heated to drive off all moisture, and dipped in a tin bath containing melted paraffin wax. The bundle should not be removed from the bath until the paraffin has begun to set, as if it is taken out too soon the melted wax will run out at the ends of the wires, and the object is to retain as much as possible in the core. For the choice of paraffin and its management see below.

III

THE PRIMARY COIL

THE writer believes it to be the general practice to wind three layers of primary in a coil of the size of that being described, and he therefore hesitates to recommend another course. There seems, however, good reason to doubt whether the third layer presents any advantage. Of this, however, the reader can judge for himself from the notes given later [1] of some experiments made to investigate this point. To enable this and other points to be tested, it is interesting and easy to wind each layer separately and independently, bringing the ends out freely, and as this relieves one from the necessity of turning back at the end of the first winding, advantage may be taken of the fact to wind each layer as a right-handed helix. The second layer thus lies in the grooves formed between consecutive turns of the first, and the third

[1] See Appendix, I., p. 145.

lies in the grooves of the second. Of course when the layers are used in series, as is generally the case, the current is led back from the end of one layer to the beginning of the next, inside the base on which the coil stands. By adopting this method, the wire as a whole is brought somewhat nearer to the core than it would be if wound in the ordinary way. In fact, it can easily be seen that the mean distance of the axis of the wire from the surface of the core in this arrangement is about 2·4 times the radius of wire, as against three times the radius in the ordinary arrangement. A suitable wire is No. 14 S.W.G. high-conductivity copper, singly covered with white silk.

It is frequently suggested that wire of square section should be used for the primary of large coils, and it is obvious geometrically that the resistance per unit of length of the primary is (theoretically) lowered by this substitution in the proportion of 4 to π, that is, the resistance of the square wire would be about ·78 of that of circular wire. But it should be remarked that the resistance of a Grove battery and leads is, unless very large cells are used, greater than that of the primary,[1] and that if the method here advocated is adopted of winding all the helices in the same direction, the

[1] The resistance of three layers of round primary will be under half an ohm, while that of six large Grove's cells in good order may be about an ohm, and of six smaller ones more than two ohms

difference between square and round wire becomes much less marked; and lastly, that square wire is difficult to obtain and much more troublesome to wind.

For the purpose of winding conveniently the core is best mounted in a lathe. This may be done as follows: A piece of wood an inch or so thick having been secured to the face plate, a hole $\frac{1}{2}$ in. deep and $1\frac{1}{16}$ in. in diameter is turned in it, and a small central hole bored right through. This is removed from the chuck, driven on to one end of the core and secured by paraffin, and serves to receive the point of the back-centre. A similar piece turned on the face plate and left there will form a cup-chuck. The strap may be (if the lathe is an American one) removed, and a winch-handle attached to the far end of the mandril. If the lathe is an English one, this will involve removing the bearings of the pointed screw against which the left-hand end of the mandril turns. But it is a great comfort not to have to pull the lathe round by the strap.

The hank of wire having then been placed on an extemporised "swift" or skeleton drum, and the end secured to the wooden cap of the core, the winding is easily proceeded with.

A simple and efficient "swift" can be made by nailing two strips of wood together a foot or so long and an inch broad (they are better "halved" together), so as to make an equal-armed cross.

This is laid on the bench, or a board, and a bradawl or French nail driven through the centre for it to turn on, a washer or something similar being interposed between the cross and the bench to diminish friction. The hank of wire is laid on the top, and four other bradawls are inserted into the arms of the cross, but not passed through the wood, one in each arm, just inside the coils of the hank of wire, to hold it in position as the wire is wound off.

It is perfectly easy to guide the wire while an assistant turns the winch, and it may be carried right up to the chuck, as there is $\frac{1}{2}$ in. of core inside the latter. If a screw-cutting lathe is available, the screw may be set at fourteen threads to the inch (the covered wire being just $\frac{1}{14}$ in. in diameter), and a sheave fastened to the tool-holder in the slide-rest to straighten the wire and facilitate winding. The end is then temporarily secured with string in every case, a foot or two being left over. The number of turns in each layer should be recorded, as this can evidently not be done afterwards.

It is very desirable not to trust entirely to the silk covering for insulation, for in such a coil as is now being described the primary spark will probably be sufficient to pierce through the insulation when two pieces of the silk-covered wire are laid in light contact. It is probable that drying and soaking in a bath of hot paraffin will enable it to resist this, but it is clearly safer not to trust entirely to the covering.

Therefore when the first layer of primary is wound, and the ends secured, a narrow white silk ribbon should be wound over it, so that each layer overlaps the last, the ends being secured by a few stitches with a needle and sewing silk. The wire before being wound should be dried and soaked in hot paraffin to saturate the silk, and the ribbon similarly treated, though possibly the treatment presently to be described will render this unnecessary. Still, the importance of keeping all insulation dry is so great, that perhaps it is on the whole better to dry and soak each portion of the winding separately before use, to avoid any possible access of moisture afterwards.

Each layer of wire is similarly covered with ribbon until all three are on, and lastly the projecting ends are tied into cables, two out of the three wires at each end being first wrapped in ribbon as an additional safeguard.

The core and primary being now complete should be carefully tested, the resistances of each layer and of the whole coil being taken and the insulation resistance between the coil and the iron core tested.

IV

THE MAIN INSULATING TUBE

THE next thing is to cover the primary with thick ebonite insulation, this insulation being the most important in the whole coil and the part subject to most strain. If the cotton insulation of the secondary, described below, be adopted, a $\frac{1}{4}$ in. of ebonite ought to be sufficient.

Ebonite tubes are generally used for this purpose. Now these are not very easily obtained; they are expensive and are apt to contain weak spots. Moreover the tube ought to fit tightly over the primary, that no space may be lost. Another method is consequently described which is cheaper and quite satisfactory.

A sheet of ebonite $\frac{1}{16}$ in. is taken, and cut (with a a tenon saw) to a rectangle about 30 in. long (or the length of procurable sheets, which is about this) and 22 in. wide. This it will be found can be wrapped round the primary coil, and will envelop it three times with about ½ in. to spare. Each of the shorter sides of the rectangle is then brought to a feather-edge with a file, the chamfer extending ½ in. from the edge. A large shallow tin bath is then constructed by turning up the edges of tin plate and soldering. Water is poured into this, and the ebonite sheet laid in the water, being kept from contact with the bottom by some strips of wood. All available heat should be then brought to bear on the bottom of the bath, as with so much evaporating surface it is not easy to boil even a comparatively small quantity of water. The means of doing this will vary according to the apparatus which happens in any case to be available. In the case of the writer two or three small petroleum stoves were pressed into service, and supplemented by all the Bunsen burners which the resources of the workshop furnished. It is one of the most convenient properties of ebonite that, if heated to a temperature not exceeding that of boiling water, it becomes quite soft and pliable, resembling leather, and, on cooling, completely recovers its ordinary condition. In dealing with it while soft no time must be lost, as it cools rapidly and hardens at once. It may be mentioned in passing that for some reason sheet ebonite, which is sold by weight, has a certain uniform price per pound for sheets of all thicknesses down to $\frac{1}{16}$ in. But the price rises suddenly for thinner sheets to something like double that of thicker ones. For this reason it is well to use the thicker sheets wherever it is possible.

It is an unfortunate thing that, so far as the writer is aware, no thoroughly satisfactory cement for ebonite has been found. Numerous inquiries were made and several experiments were tried by him with indiarubber dissolved in benzine, gutta-percha in the same, and in carbon disulphide, melted shellac, and various combinations of the latter with resin and beeswax; but none of these cements made a satisfactory joint. The method which is recommended, therefore, is to take the sheet of ebonite out of the hot bath, wipe it rapidly with a hot towel, and roll it round the primary coil, tying it up temporarily with tape. On removing the tape afterwards it will probably be found that the ebonite sheet has some resilience and springs open a little. This may be due to the inevitable loss of time while wiping and rolling it, enabling the sheet to resume to some extent its elastic character by cooling. At the present stage this is, however, no disadvantage, and the way this tendency can be overcome when it would be objectionable will be mentioned presently. Indiarubber solution may be brushed in as far as can be reached between the layers, and the whole tube carefully served with a wrapping of narrow white silk ribbon, previously dried and paraffined. The ribbon serves to compress the spirals of the tube and keep it close to the primary, at the same time adding its modicum of insulating power.

Next a tin bath should be soldered up the exact breadth of the finished tube and an inch or two longer. The bottom should be rounded to fit the tube exactly. This is easily done by taking a sheet of tin plate about 2 ft. long and 9 in. or 10 in. wide, and bending it in to the shape of the letter **U**, keep-

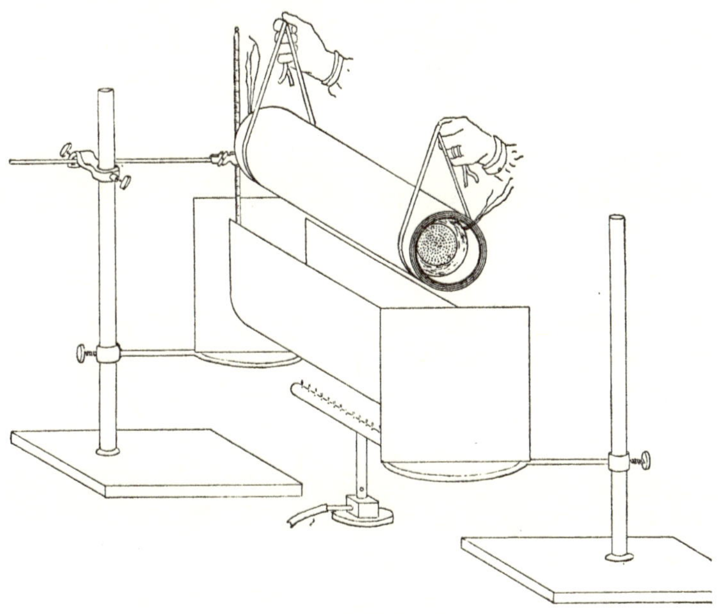

FIG. 1.—Completing Insulation of Primary Coil.

ing the rounded bottom to the radius of the tube, and then soldering squares of tin plate to the ends. Paraffin is then melted in it by placing it over a "rampe" burner turned very low. A thermometer is inserted and the bath carefully kept at about 120° C. The primary and its tube are laid in this, and allowed to soak in the melted paraffin (see Fig. 1). Every few minutes one end or the other is lifted up (not, however, above the surface) by tapes left under the tube for that purpose, and bubbles are allowed to escape from the crevices between the primary coils, the ebonite wrappings, &c. This should be continued until bubbles entirely cease to appear when the ends are lifted. Then the gas may be turned out and the whole apparatus left for the night to cool and harden.

This having taken place, the coil will of course be found to be tightly stuck into the bath by paraffin wax, and the next thing is to release it. This is simply effected by playing on the outside of the bath with a Bunsen burner, so as to soften the wax superficially and allow the contents to be withdrawn. They, of course, form a casting of the inside of the bath, and require trimming with a spatula until the tube is exposed, and the masses of wax projecting from the ends cut off straight and flush with the ends of the tube. It must be borne in mind that paraffin contracts very considerably on cooling, and the only way to fill any hollow with it is to adopt some such method as that just described, by which a superfluity of wax is used and removed with a knife when set. The temperature at which the hardest paraffins melt is, of course, much lower than that at which the bath is kept, but by keeping everything at a temperature ten or fifteen degrees above the boiling point of water, there is reason to hope that any moisture which may still linger either in the core, the primary, or the tube, may be effectually got rid of, while at the same time over-heating and decomposition of the wax are avoided.

The tube will now be $\frac{3}{16}$ in. thick in the walls, but it is well to have it $\frac{1}{4}$ in. thick. A new strip is accordingly cut, long enough to go once round the tube as already made, with an inch to spare for feather-edging. This is heated as before and rolled round a brass tube *smaller than the finished ebonite tube*, so that when cold it may require to be sprung open and will embrace its contents tightly. It may be mentioned that to secure the complete rolling up of the ebonite before cooling has begun, it is convenient that the brass tube on which it is to be wound should be fitted with ends and connections, and heated internally by steam from a tin boiler. This, however, is a refinement which is probably unnecessary. Of course throughout these manipulations thick gloves should be worn, but, even so, the heat to the hands is not very pleasant.

When the last tube is set, it is sprung open, and kept so by wooden wedges, and the primary, in its spiral ebonite wrapping, smeared with indiarubber solution, slipped in, and the wedges removed. The outer tube will then close in upon the spiral, and embrace it tightly. The core and primary are now complete and insulated. The insulation should then be tested by connecting the primary wires to one pole of an induction coil, and endeavouring to break down the tube by sparking. The largest coil which can be made available for this purpose should be used, up to a 6-in. spark or so.

If this is unattainable the test must of course be omitted, and the maker must trust to his good work to have produced a satisfactory result. But if it is possible to make the test, it should on no account be neglected.

V

THE CONDENSER

THIS is the simplest part of the coil, and all the handbooks contain instructions for making it. But the method adopted by the writer seems to be an improvement and worth describing.

The ordinary and well-known method is to cut strips of tinfoil one or two inches less in width and length than the dielectric used to separate them—generally paraffined Bank post paper — to form these into a pile; and when the pile is completed to put some heavy weights on the top which (as one author says) "may be increased to half a ton, and the efficiency of the pile increased thereby." The condenser is then bound up with tape between two boards.

Now on this method it must be remarked that the capacity of a condenser is expressed by the formula—

$$K\frac{A}{d}$$

where K is a constant depending on the specific

inductive capacity of the dielectric, and A and *d* are respectively the total area of the tinfoil, and the distance between one sheet and the next, so that the capacity depends directly on the nearness of the plates to each other, and the high value of the dielectric constant for the material separating them. Any method therefore that will exclude air from between the plates (the specific inductive capacity of air being 1 and of paraffin about 1·9) and at the same time bring the plates closer together will increase the efficiency of the condenser. This is the object of the weight which "may be increased to half a ton,"—if the resources of the workshop include a Bramah press, a piece of apparatus not often found in private laboratories.

But the following method is, it is submitted, better, and requires no elaborate apparatus. A number of sheets of tinfoil 15½ in. × 9 in. are cut, and half of these are soldered to one strip of copper and the other half to another, so that they resemble two books with copper strips for backs. About 70 or 80 sheets will be found sufficient. About 150 or 170 sheets of Bank post paper are cut, about 15½ in. × 10 in., and inspected for pinholes, the number being double that of the foils, with a few over, and, a bath of paraffin having been provided, the sheets of paper are each separately held to the fire till quite dry and hot, and immediately plunged into the bath of hot wax. As each is removed from the bath it is allowed to drain by the corner and laid upon a sheet of tin plate as it cools.

Too much stress cannot be laid upon the importance of thoroughly drying all paper, cotton, &c., which is to be paraffined, and used for insulating purposes. An experiment made for the purpose of testing this showed that a sample of paper, apparently quite dry to the touch, paraffined in a bath, was broken down by a potential difference capable of producing a spark 0·1 in. long in air; while an exactly similar paper, thoroughly dried at the fire and then immersed in the same bath, was equivalent to 0·55 in. of air in dielectric strength.

The toasting should not be enough to brown the paper, though on one or two occasions when it was carried to the pitch of slightly discolouring it, no deterioration in dielectric strength could be detected.

Not more than $\frac{1}{4}$ or $\frac{3}{8}$ of an inch margin need be left round the tinfoil sheets, as in the method being described the whole condenser becomes a solid block, and the dielectric strength at the margins is much greater than on the flat of the sheet.

The books of tinfoil are laid on the table with their backs towards each other in the usual way, with three or four paper sheets in a pile between them laid on a thin board the same size as the paper sheets, and the sheets of tinfoil turned over from each book alternately, two sheets of paper being interposed between each pair of consecutive foils (see Fig. 2). An experiment showed that a single sheet could be pierced by the primary spark. When the last sheet of foil has been turned down, three or four papers are laid on the top. As the margin left is narrower than is usually recommended, care must be taken to lay all the sheets centrally and evenly.

Finally a thin board, similar to the former one, is laid on the top of the pile.

A special bath is prepared in which the pile may be warmed while being gently pressed down by

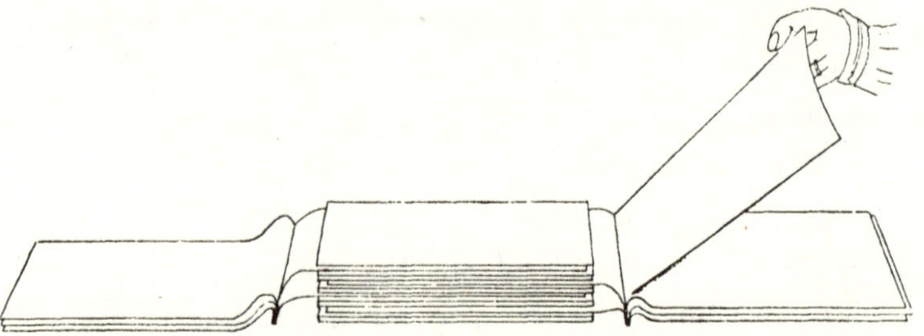

FIG. 2.—Building up the Condenser.

moderate weights. This may consist of two tin trays of the same width as the pile and an inch or two longer, soldered together after the manner of a hot-water plate (Fig. 3). To strengthen the tin bottom, which might, if only supported by the soldering at the edges, be crushed in by the weight, it should be arranged that the distance between the bottoms of the two trays is equal to the external diameter of some compo gas-piping, and a length of this is bent up into a pattern and laid between the two bottoms, the ends being brought out through the side of the under tray. This pipe serves the double object of supporting the false bottom and heating the bath by steam from a tin boiler. The bath should have also a waste spout soldered in, about half an inch above the false bottom (Fig. 3, A, B, C).

Things being so arranged, the condenser is laid

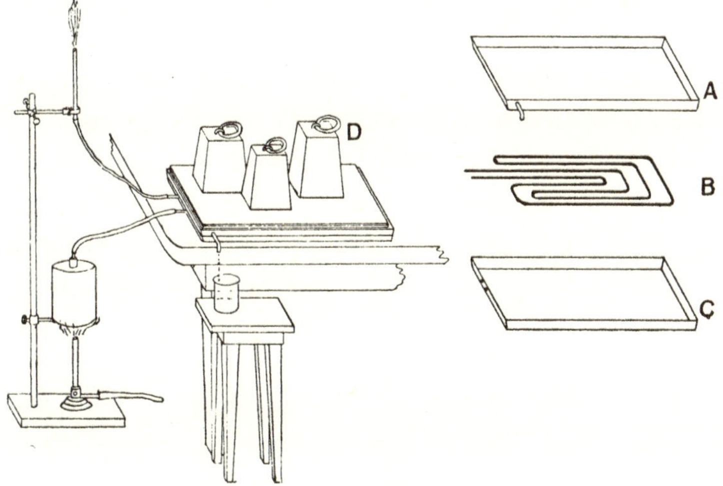

FIG. 3.—Treatment of the Condenser to Increase its Capacity.

in the bath, and some moderate weights—20 or 30 lbs.—placed on the top. A large Bunsen burner being placed under the boiler, the bath is soon brought to a temperature sufficient to melt the paraffin. Heat is continued until the whole condenser is melted and squeezed down by the weights. Its thickness will thus be considerably reduced, and its electrical capacity consequently very much increased.

An experiment made with the object of testing the value of this suggestion gave the following results. Four condensers of different numbers of sheets were made up, and their capacities measured before and after re-heating. The capacities before heating were respectively ·122, ·054, ·066, and ·033 microfarad. Afterwards they were ·84, ·367, ·449, and ·22 microfarad. It will be seen that the capacity was increased in the ratio of 6·8 to 1 about. So that the process made them as efficient as they would have been with nearly seven times as much tinfoil if they had not been re-heated.

While the heating progresses, the superfluous paraffin runs out through the spout and is caught in a vessel placed for the purpose (Fig 3, D). When the whole condenser is heated throughout, the gas is extinguished and all allowed to cool. Thus the sheets and the two boards are all stuck together, and the condenser is surrounded by a solid wax edging completely filling and adhering to the bath. A rapid superficial heating of the bath serves then to dislodge the contents, which only require tying together with string as an extra—though hardly necessary—precaution.

If more convenient, the steam boiler may be dispensed with, and the condenser, on its double dish, laid on the hot-plate of a kitchen stove ; but if this method be adopted, arrangements should be made that the interval between the false and true bottoms of the bath should be filled with water to avoid overheating. Care must also of course be taken that the wax as it melts drips clear of the fire.

The capacity may, if desired, be measured (see Appendix II.). A suitable capacity for a coil of the size now being described would be about half a microfarad.

It is pleasant to be able to vary the capacity of the condenser, for the purpose amongst other things of studying its effect on the working of the coil. This, if desired, can be very easily done. Instead of making up two books of foils containing 40 sheets each, we make up two pairs, one pair of 25 sheets each and the other of 15. They are soldered to separate copper strips insulated by being kept from touching. This gives us four different condensers—(1) one of 15 pairs of foils, (2) one of 25, (3) one composed of the two sets in "cascade," and (4) the full condenser in parallel. The ends can be brought out to separate binding-screws and connected in the appropriate way by copper or brass strips. This may be according to the diagram of connections under baseboard on page 111.

VI

THE COMMUTATOR

THE construction of the ordinary drum-commutator is well known, and will at once be evident from the diagram Fig. 4. A small drum or cylinder is turned out of some insulating substance such as ebonite, ivory, or boxwood, and has attached to it two pieces of brass of the form shown in Fig. 4, A.

These are attached to the drum so that they form end-journals and rubbing pieces in one. Care must be taken that the screws by which they are attached to the drum do not short-circuit them (see Fig 4, B), The trunnions turn in little standards fastened to the baseboard, which form bearings for the drum, and one of the trunnions is made long enough to carry a cross-handle by which the drum is revolved. Two upright springs are also attached to pieces screwed to the baseboard, so that when the drum is in one position the springs are pressing one on each of the rubbing pieces, and when the drum is turned through 180° the springs are again so pressing; but this time each spring presses on that rubbing piece with which at first the other spring was in contact. Wires run under the baseboard from the two springs to the battery terminals, and wires from the two bearings run respectively to one of the pillars of the break, and

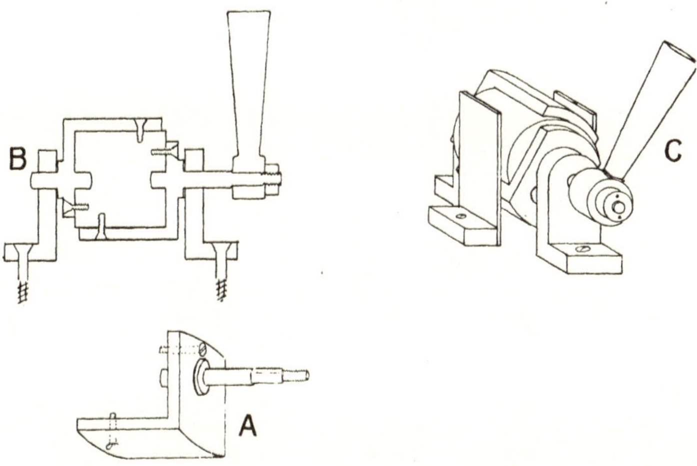

FIG. 4.—Details of the Drum-commutator. A, rubbing piece with spindle, showing attachment for handle; B, general section; C, view in projection of complete commutator.

one end of the primary wire. Fig. 4, C shows the commutator set up.

This commutator works well and cannot be much improved on. Possibly, however, the considerations which led to a variation of it, which will now be described, are worth attention.

It is, of course, desirable to diminish in every

possible way the resistance of the primary circuit. Now, in the ordinary commutator there are four joints where resistance may be found, viz., at the two bearings of the cylinder, and at the two springs. These may be reduced to two, and the contact

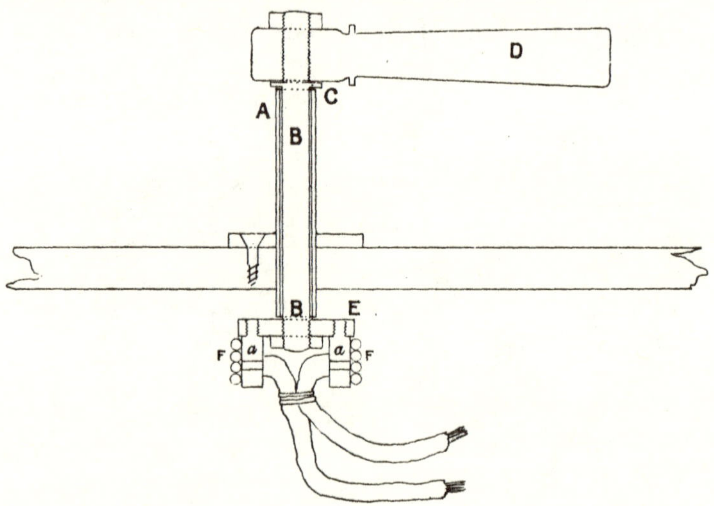

Fig. 5.—Modified Commutator attached to Baseboard of Coil. A, fixed tube; B, spindle; D, handle; E, ebonite disc; *a a*, brass pins; F F, springy brass wires on which *a a* rub to make contact.

much improved. The design is shown in section in Fig. 5.

A is a piece of brass tube, soldered into a flange and screwed to the baseboard; B B, a brass rod passing easily through the tube, and supported on it by a shoulder C. At the top is fastened by a nut an ebonite handle D, and at the bottom is screwed on a disc of ebonite E. Into this disc are screwed two brass pins, *a a*, about $\frac{3}{16}$ inch in diameter, and having the ends of stout flexible insulated wires passed through holes in them, soldered, and cut off flush. These flexible wires can now be procured much more easily than used to be the case, since the advance of electric lighting has produced a market for flexible leads of low resistance. They are composed of a large number of No. 40 copper wires laid together and covered with indiarubber and cotton. Instead of bearing on flat springs as is usual, each pin bears on a spring made by laying half a dozen straight brass wires F (sold for use in musical instruments) parallel and touching each other, and soldered at one end to a piece of brass (not shown in the figure). To these pieces stout wires are soldered to make the connections underneath the baseboard. The flexible wires are soldered to washers gripped by nuts under the battery terminals.

If the ordinary commutator is examined, generally only a small part of each spring shows by its polish that it makes contact with the drum, whereas every one of the brass pins makes good contact.

The parts of these springs where the moving pins rub should be flattened with a file, and, with a rat-tail file, a slight groove should be made across each spring, in order to guard against the possibility of its pressure dislodging the rotating part of the commutator from its position when the coil is "on." Moreover, the grooves enable any one turning the handle to feel when the commutator is in the right position. Of course the handle must not be turned right round, as this would twist up the flexible leads ; but this danger is easily provided against. Unless the baseboard is unusually large, the handle of the commutator may be made long enough to secure that it shall come in contact with the break pillar if an attempt is made to turn it right round. If, however, in the case of a coil with a larger baseboard this were not the case, it would be well to screw into the baseboard a small ebonite post, to render a complete revolution of the handle impossible.

VII

THE BREAK

FOR ordinary purposes the Apps break is highly convenient and not easily to be improved upon. But a mercury break is also very desirable, and a hand break giving single contacts only is very agreeable for experiments with Space-telegraphy, Hertz-waves, &c.

It is consequently a desideratum to be able to change readily from one break to another, and the following method is recommended with this object.

The breaks are all mounted on separate mahogany baseboards, $6\frac{3}{8}$ inches square. Along the bottom of each board are let in two strips of brass $\frac{3}{4}$ inch wide and $8\frac{3}{8}$ inches long. Near each end of each brass strip is made a slotted hole or notch, care being taken that the two notches on each strip are exactly the same distance apart. Each strip projects one inch beyond the side of the wooden base (see Fig. 6).

Four screwed studs project from the main baseboard of the coil in such positions that the notches in the brass strips slip easily over them, and milled nuts can be screwed down tightly over them, firmly

FIG. 6.—Baseboard for Removable Breaks.

securing the break to the coil. These studs and nuts also form the electrical connections for the break, the two on one side being in connection under the baseboard with the condenser (or if a variable condenser be used, with terminals as shown in Fig. 26), and the other two with an end of the primary and a lead from the commutator respec-

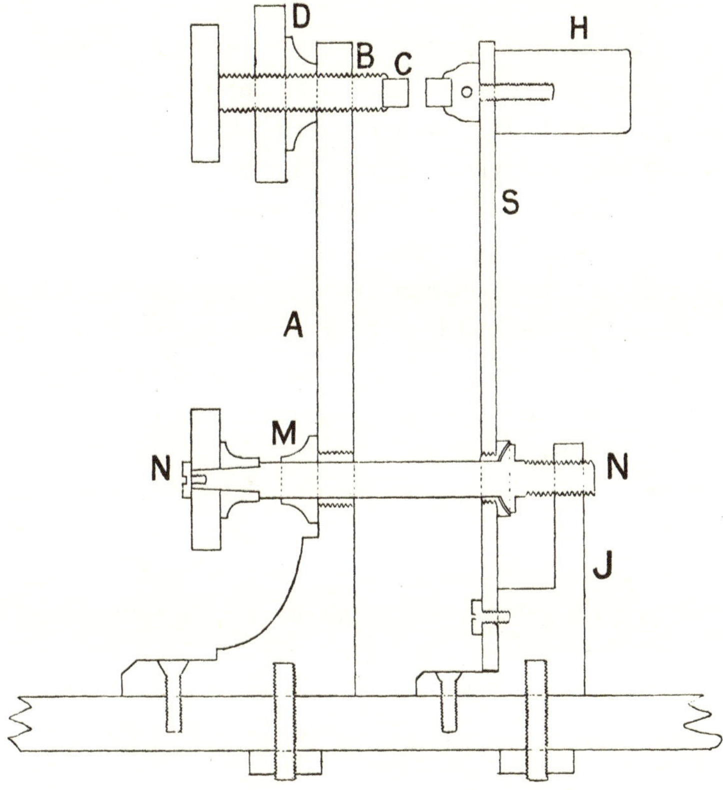

FIG. 7.—Vertical Section of the Apps Break.

tively. The strips are, of course, soldered to the pillars of the break—if it is an Apps—or in the case of a mercury break, to wires leading to the cups.

By a single turn of the four milled heads the break is entirely disconnected and can be removed from the coil and another substituted ; the change taking a few seconds only and all contacts being good.

The Apps break is so well known that a detailed description is only inserted for completeness.

A, Fig. 7, is a standard firmly[1] secured to the baseboard carrying a screw B, armed with a platinum point C, and furnished with a check-nut D. It is of such a height that when it is screwed to the baseboard, the hammer faces the ends of the core-wires. If the coil is to be used seriously the maker is urged not to be too economical with the platinum. It is unfortunate that this indispensable metal is now so costly, but if a large coil is much used—especially if used at full power—the platinums burn away rapidly ; $\frac{3}{16}$ inch is the smallest-sized rod that can be recommended, and $\frac{1}{4}$ inch is better. It is manifestly extravagant to thread the platinums and screw them into the ends of the contact and hammer-head screws respectively. If the ends of the platinums and of the screws are nicely flattened,

[1] It has always been considered that the pillar supporting the contact screw cannot be too rigidly fastened to the baseboard. Lately, however (*Journal de Physique*, vol. vii., 3rd series, June 1898, p. 342), M. Izarn has maintained that a certain elasticity is an advantage, and he mounts the pillar on an elastic base. The analogy with the movement of an ordintry electric bell will strike the reader. But it would seem well to wait for confirmation of this conclusion by other experimenters.

they can be securely attached by silver solder and no platinum need be wasted. The platinums never get hot enough in working to soften silver solder. S is a piece of "hard" brass ribbon, $\frac{1}{16}$ inch thick and about $\frac{1}{2}$ inch or $\frac{5}{8}$ inch broad, to which the hammer-head of soft iron is fastened by a screw, and which is attached to the baseboard by a short standard J. N N is a screw working through J, and passing through an insulating collar M in A. By unscrewing N N the spring S is brought to bear harder against the contact screw, and the magnetism of the core is allowed to increase to a greater degree before contact is broken.

It will be seen that with this construction, either the iron core of the coil must be long enough to reach nearly to the end of the ebonite tube, or else a "pole-piece" of soft iron must be attached to the ends of the core wires. This has never seemed desirable to the writer, as the pole-piece cannot be distinguished, magnetically, from the core itself, and should, to avoid eddy currents, be itself composed of wires. This is, of course, tantamount to lengthening the core, which cannot be done without altering the distribution of the lines of induction. And when the spring break is discarded for a mercury or hand break and the coil is worked at its full power, it is well that the path which a spark must take to pass from the secondary terminal to the core should be as long as possible. For these reasons it is suggested that a stiff brass rod should be inserted between the spring and the hammer-head to enable the hammer to reach the core without unduly lengthening the latter. The paraffin filling up the end of the ebonite tube where the break is to come is carefully bored through until the ends of the core wires are reached. A heated

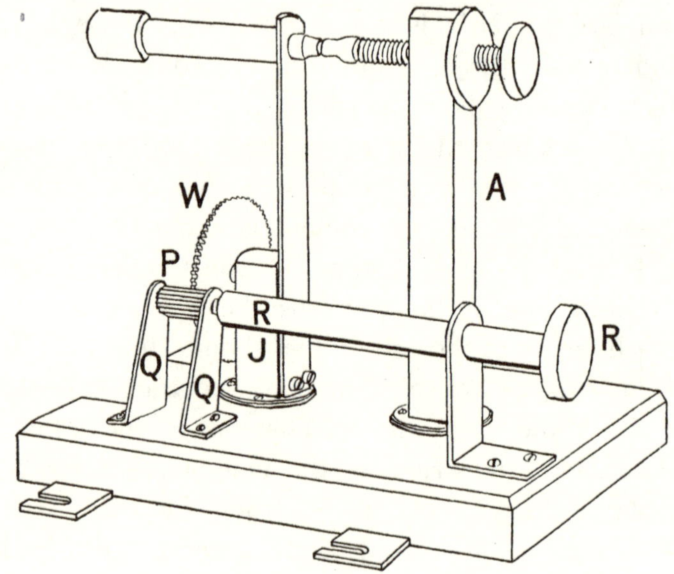

Fig. 8.—Modification of Apps Break.

iron rod to begin with, and a gouge to finish, do this easily.

The following little variation in the Apps break is suggested *quantum valeat.* It is not much more trouble to make, works well, and is pleasanter to use (see Fig. 8). Instead of piercing the spring to admit the screw N (an operation which must, of course, weaken the spring slightly), the screw which is threaded through the piece J terminates where it bears on the spring with a rounded end. The stalk of the screw and the ebonite milled head are omitted, and no hole is made through the standard A. But on the end of the screw remote from the spring an ordinary toothed-wheel W is soldered, gearing with a pinion P. This pinion (which is long enough to remain always in gear with the wheel in spite of the travel of the screw) is held in two brass bearings Q Q, and its steel axis is prolonged by an ebonite rod R R, ending in a milled head of the same material.

By this construction, to tighten the spring the head is turned in the direction to which one is accustomed in tightening anything, viz., the right-handed direction (instead of the contrary as in the ordinary form); also much less force is required, and the long ebonite shaft puts the operator well out of reach of accidental primary shocks, to which he is rather liable when feeling in the dark for the screw to adjust the ordinary break.

VIII

THE SECONDARY COIL

REVIEWING progress so far made, it may be considered that the primary coil is mounted on its temporary stand, under which the condenser is attached by wooden strips and screws, or in any other convenient rough and ready way, that the break studs and commutator are fitted in their correct relative positions, and that flexible leads have been attached to the temporary stand by which contact with the primary can be made whenever desired. On attaching break and battery the coil ought to give all the effects, such as lighting incandescent lamps,[1] giving shocks, &c., which depend on the condenser and the self-induction of the primary. We are, then, now ready for the most important process, viz., the construction of the Secondary Coil.

If thought desirable, the middle of the secondary coil may easily be made accessible, so that half of

[1] Fleming, "The Alternate Current Transformer," new edition, vol. ii., p. 113.

it can be used without the other, or both at once, for the purpose of obtaining two simultaneous sparks. This is of course a minor point, but gives practically no more trouble, and the construction also has the advantage that by it the great weight of the bobbin is supported at the centre as well as at the ends. If this is desired it may easily be effected by adjusting on the main ebonite tube a central diaphragm or partition of $\frac{3}{16}$ ebonite similar to one of the ends, but thinner, to which two binding-screws are attached, one connected by a brass strip with the inner end of one half of the secondary and the other with that of the other one. In ordinary work with the coil, these two binding-screws are connected by a short straight wire with little screwed knobs (principally for ornament) on the ends (see Figs. 23 and 24, pp. 103, 107).

The partition gives the finished coil an appearance similar to those wound in two sections, but of course the central partition is in this case not relied on to assist insulation at all.

In the main the method of winding the secondary wire described by "Inductorium" (see the *English Mechanic*, vol. xxi., p. 450) cannot be improved on. But the modification about to be described is, it is suggested, not without importance, as it avoids the troublesome inside solder-joint and gives a cheap and satisfactory means of obtaining high insulation at the places where it is most needed.

It is first necessary to describe in some detail the machine used for winding, which has to be home-made, as no such machine is, so far as the writer is aware, to be found in any instrument-maker's catalogue.

Fig. 9 is a vertical section of the machine (which is also seen in perspective in Fig. 11, p. 59). B A B in Fig. 9 is a cast-iron stand consisting of a base A and two standards B B. These are drilled and tapped to receive the centres C C, which had better be of steel, and must fit easily, as they are constantly being taken in and out while the machine is being used. For the same reason they should carry check-nuts to tighten them in position.

Between these centres revolves the spindle D, having countersunk ends e e′, and carrying (if desired) a threaded worm, f, to work a simple revolution counter; g is a disc soldered on against a shoulder; h is a parallel part, $\frac{7}{16}$ inch in diameter; and j is threaded to take a milled nut.

Over the parallel portion of this spindle drop with a nice fit two planished brass plates P P′ 8 inches in diameter, and a smaller and thicker plate or disc, seen between them at Q, $3\frac{1}{4}$ inches in diameter and $\frac{3}{16}$ inch thick. The edge of this last disc is slightly coned, and there is a third planished plate P″, turned with a central hole just greater than the largest diameter of the disc Q, and capable of being attached to P′ by two small screws S S′.

The plates P, P′, and P″ have their edges turned

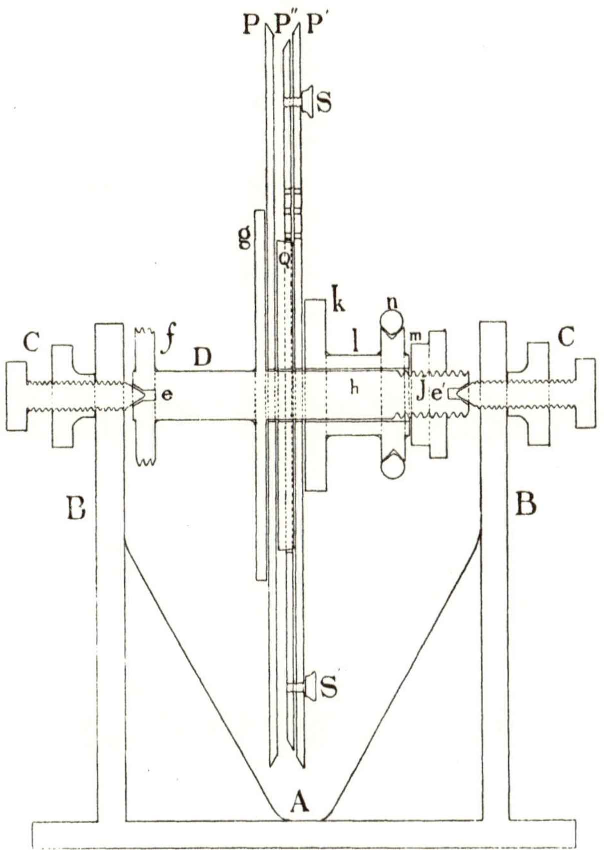

FIG. 9.—Vertical Section of Winding Machine for Sections of the Secondary Coil. A B B, casting supporting C C, centres on which the spindle D h turns; P P′ P″, circular plates between which the section is wound; S S′, screws holding P″ to P′; Q, thickness piece; k l m, piece sliding on D h and holding P P′ against the fixed flange g by nut m; n, rubber ring by which motion is transmitted from rim of flywheel to the whole winder: f, screw to drive counting apparatus.

conical so as to lead the wire smoothly into the groove.

These plates are held firmly on the spindle by the piece k l m.

This piece may either be made from a casting, or built up from discs and tube. In either case it must be an easy fit over h. k is a disc matching g, the two holding between them the planished plates and compelling them to run true; m is a thick disc grooved at the edge to receive an india-rubber umbrella ring n, which by friction from the edge of a flywheel drives the whole apparatus. As in the course of winding it will be necessary to mount and dismount the spindle many times, it will be seen that this frictional gearing is very much more convenient than a driving-band which has generally been used for the purpose.

When the plates with their disc and the piece k l m are slipped on to the spindle and the whole tightened up by a milled nut on j, the entire arrangement should be able to turn quite true between the centres, and with very little friction.

A counter is rather a luxury than a necessity, but probably few will dispense with it, as it affords much information about the coil when complete. It is shown partly in section in Fig. 10, and consists simply of two toothed wheels a b, f y, carrying respectively 100 and 101 teeth. The wheel with 101 teeth is mounted on a hollow spindle which is squared in front to take a short pointer, p, just as is the case with the largest of the dial-wheels in a clock, the one whose spindle carries the hour-hand. The 100-wheel rides loose on this spindle, and every tenth tooth is numbered.

The hollow spindle rides on a stud S, pro-

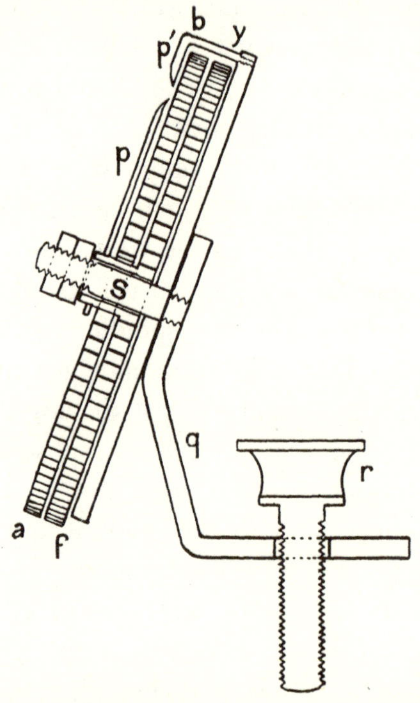

Fig. 10.—Counting Apparatus. a, f, wheels having 101 and 100 teeth respectively; a, having the teeth numbered on the face, and reading units up to 100 against pointer p′ and hundreds on pointer p, which latter is keyed to f.

jecting from the face of a small cock or bracket q, which can be secured by a milled-headed clamping-screw r to a piece attached for the purpose to the left-hand standard B (Fig. 9), so that both toothed wheels gear at once with the worm f, (Fig. 9). The cock on which the wheels ride carries a little pointer at the top p′, projecting over the wheels and reading on the front of the 100-wheel. It will be seen that when the shaft of the winding machine has made a hundred revolutions, the zero of the wheel a b has just come round again under the pointer p′. But the wheel f y, which has also advanced 100 teeth, has not quite made a complete revolution, as it has still one tooth left. This causes the hand p to lag behind, and read one on the wheel a b. Thus if we call the readings of p′ units and of p hundreds, the counter gives exact readings up to ten thousand. It can be set to zero in a moment by unclamping r, and giving a slight turn to the cock q. This disengages the wheels a, b, from the worm, and they become independent again.

It will be noticed in the winding machine that the distance between the plates P P′ (Fig. 9) being $\frac{3}{16}$ inch, and the thickness of the plate P″ $\frac{3}{32}$ inch, we have an available groove between the plates $\frac{3}{32}$ inch in thickness when P″ is in position, and double this thickness when P″ is removed.

It remains to describe a few accessories to the winding machine, essential to its working. These are the paraffin-trough, the stand for reels of wire (which we may call from its occasional use the "unwinder"), and the "wiper," a small, but, as experience showed, a very necessary adjunct.

Fig. 11 shows a perspective view of the whole winding apparatus ready for use. A is the machine already described. B a flywheel mounted on a little standard and carrying a winch-handle. This standard should be attached to a small independent baseboard having slotted holes in it through which pass screws firmly attached to the main baseboard, and having wing-nuts working on them, n n. The position of the flywheel can then be accurately adjusted so that it bears with just sufficient friction on the rubber-tyred wheel on the winding-spindle, when it can be clamped by the wing-nuts.

w is the wiper, which will be described presently.

C C′ is a trough made by bending up and soldering tin plate. It is stiffened by cross-braces of wire, D D′, and has a small pin-hole gas-burner underneath it, E. Wire hooks are soldered to the bottom inside to guide the secondary wire below the surface of the melted wax, with which it is filled, and the ends are smoothly turned over so as to present no roughness which might abrade the silk. The wire horns, h h′, seen at the top at one end, are to prevent the wire slipping off laterally. It has no tendency to do so at the other end.

F is a spindle made of a piece of $\frac{1}{4}$ inch brass rod, having a small grooved sheave soldered near one end, between which and a nut at the other end the reel of wire may be clamped. The spindle carrying the reel is easily taken in and out of the uprights between which it turns, by pulling these

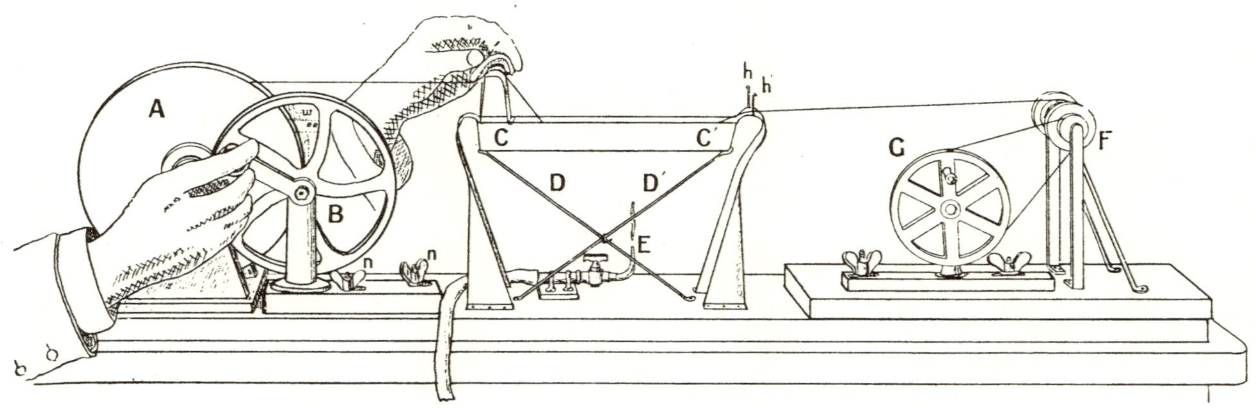

FIG. 11.—Winding Apparatus complete. A, winding machine driven by flywheel B, to which wire passes from reel F through melted paraffin in trough C C′ ; n n, nuts holding down board to which standard of flywheel is fastened ; E, gas flame keeping paraffin melted ; G, wheel used occasionally for unwinding.

latter slightly apart. They are made springy enough to admit of this. The little sheave drives the unwinding wheel G by a thread, but ordinarily of course it drives G instead of being driven by it. Of course if preferred the connection between this spindle and its driving-wheel may be frictional, and similar to that of the winder; or the driving-wheel may be omitted entirely, and, if unwinding becomes necessary, the reel of wire may be turned by hand. But the convenience of having a ready means of quickly unwinding when necessary is very great.

There is a fitting which was never attached to this apparatus as used by the writer, but which would, it is believed, have been of advantage. If the wire breaks—and an occasional break is inevitable—the reel F continues to turn at a great pace, while the wire which it pours out is no longer wound up. This causes a troublesome tangling between F and C′. Occasionally this tangling is so great that the whole reel of wire becomes useless. In the writer's case this was not quite so serious, as he did not work alone, and when a break occurred a watchful hand was instantly laid on the reel F and the damage much mitigated. But as it must often happen that such assistance is not forthcoming, and as there is really no reason why one person should not conduct the operation of winding by himself, it occurs to the writer that if between F and C′ the wire were to pass under a wire hook attached to the long end of a light lever, the tension of the wire keeping the hook up might so keep a break-block off the reel F, which would be instantly applied when the tension of the wire between F and C′ was relaxed. The writer has, however, given no diagram of this, nothing being figured or described in this book which has not stood the test of actual experience.

Such being the description of the winding apparatus, we may consider the method of using it. And first the materials must be at hand. These are, according to the writer's method—

1. Insulating paper.
2. Sewing cotton.
3. Wire.

We will consider these in order.

Such descriptions of coil-winding as the writer was able to get from the meagre sources available seem to imply that the sections cannot be adequately insulated with less than four, six, or even a dozen thicknesses of paper. And these are generally applied to the section of wire after winding, thus imposing upon the maker the troublesome task of manipulating a section with no other support to keep it from collapse than the adhesion of the wax.

But experiment showed that if the paper were of the right quality and properly dried before paraffining, one thickness was sufficient to resist a greater spark than the section was capable of producing. Nevertheless, to increase the factor of safety the paper was doubled where the difference of potential on opposite sides of it was greatest.

We will therefore suppose that the maker has prepared a sufficient number of discs of paper 8 in. in diameter and with a central hole 3¼ in. in diameter. The paper recommended is fair quality white blotting-paper ; that used by the writer was marked "Ford 428 Mill," and was not adopted until several other sorts of paper had been experimented with. It takes up plenty of paraffin and is but slightly (if at all) sized. Other discs or rings should be cut of the same external diameter, but having a central hole 6 in. in diameter, leaving only about 1 in. as the width of the ring. The waste from this last cutting will enable a third set of discs to be cut having external diameter 5½ in. and diameter of hole the same as in the first set, viz., 3¼ in. If it is thought worth while, circular templates may be cut out of brass or sheet zinc and have their edges smoothed in the lathe, when the discs can be nicely cut by laying the paper on a sheet of glass and the template on the top, and running a penknife round. But it takes very little longer to mark the discs with the compasses and then follow the circular lines with a pair of scissors.

It will be seen that by laying one of the first set on one of the second we have an insulating disc, having more dielectric strength in the parts near the edge than in those near the centre; while by laying one of the first on one of the third set, the dielectric strength is greater near the central hole. (See Figs. 12 and 13, in which however, of course, the thickness is much exaggerated.)

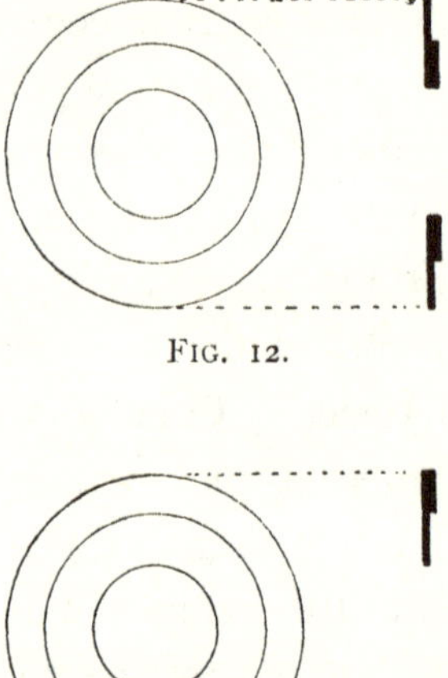

FIG. 12.

FIG. 13.

Insulating Papers for Sections of Secondary.

The discs are all to be held to the fire one by one until thoroughly dry, and plunged while hot into paraffin kept melted in a tin baking-dish over a small gas-flame.

This being perhaps the most vital point in the whole coil, it is advisable to consider in some detail a few of those small points on which success may depend. First, as the fingers cannot be used to hold the paper rings to the fire, some other method must be adopted. The most obvious method is to use a pair of crucible tongs. But a moment's reflection will show that wherever the paper is gripped in the tongs, the cold metal will protect it from the radiation of the fire, and a spot of imperfect drying will result. This may be mitigated by shifting the tongs from place to place while the drying process is continued, but a simpler way is to hang the paper ring over a wire hook—or the crucible tongs themselves used as a hook—passed through the central hole. The paper may be removed from the fire a few moments after visible steam has ceased to come from it, and before it begins to char; and it should be then immediately plunged into the paraffin bath. A disc of the first set having been so treated, one of the second should be dried and plunged in on the top of the first. They can then be laid in position and so withdrawn from the bath, when they will cool and adhere together. Then one of the first set followed by one of the third may be treated, and caused to adhere. Finally we shall have a number of papers which can be sorted into two sets. They are all of 8 in. external and 3¼ in. internal diameter, but one set will be thickened on the outer, or peripheral portion, and the other on the inner portion (Figs. 12 and 13).

It is now extremely desirable to form some idea of the resisting power of the discs to sparking. For this purpose a small coil is taken—or an influence machine would no doubt answer, but a coil is to be preferred--and the discharging points set to about $\frac{5}{8}$ or $\frac{3}{4}$ in. spark. While so set, a wire

FIG. 14.—Instrument for Testing Insulating Papers.

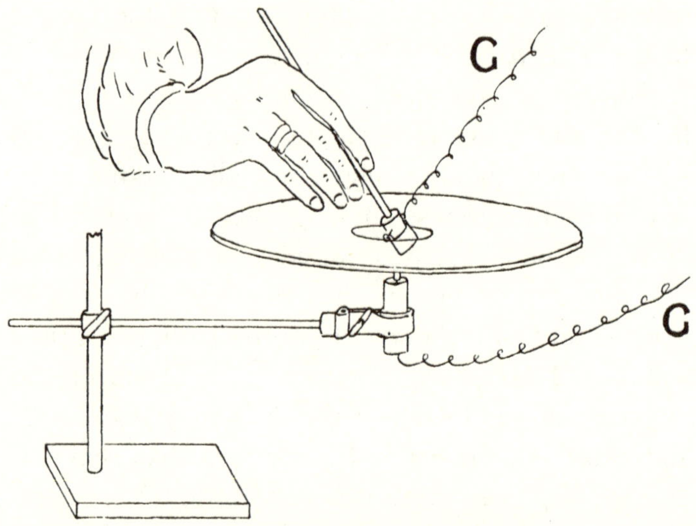

FIG. 15.—The same in use.

should be led from one to a metal plate lying flat on an insulating support, and the other to a little arrangement with a glass handle and a wire frame, by which the current can be taken to every part of the upper surface of the paper without fear of scratching it. Figs. 14 and 15 will show what is meant. A is a glass rod, B a cork stuck on the end, C D a thin brass wire thrust through the cork and having a little ring turned up at each end. E F is a little wire frame bent as shown and articulated to C D at C and D so as to hang loosely. G G (Fig. 15) are wires leading to the coil. It expedites the testing if it can be arranged that the metal plate on which the paper disc is laid shall be mounted on a pivot, so that it can be set spinning before the coil is turned on. Then by simply holding the glass handle so that E F lies gently on the paper, a test is made in a moment. But if it is not desired to take the trouble to arrange a revolving table, then the little testing electrode just described must be drawn over the section like a garden rake as the section lies upon the plate. In either case the latter should be circular and about ½ in. less in diameter than the paper disc, or the portions of the latter near the edge do not get properly tested. A percentage of papers are sure to get pierced, and should of course be rejected. But if the drying and paraffining have been well done, this percentage should be a small one.

It might be thought that the papers could be easily tested by placing them between two metal discs connected simply to the secondary terminals of a small coil. But it must be remembered that such an arrangement constitutes a condenser, and if the coil is so connected it will be found that unless a large one is used the length of spark is so much reduced that no test can be made.

The other insulating material is sewing cotton. It is probably quite unimportant what cotton is used; the writer used No. 26 of a brand called "Boar's Head." But dryness of the cotton is of the first importance, as likewise of the silk covering of the wire. To ensure this, a sufficient number of reels both of wire and cotton should be first thoroughly dried at the fire, and then kept till wanted under an air-pump receiver (standing on a flat plate luted as usual with grease) in company with a large saucer of calcium chloride. It is not suggested that they should be kept in vacuo, which would be an unnecessary refinement, but merely that having once for all been thoroughly dried, they should not again be exposed to any atmosphere but a perfectly dry one until required for use.

The object of the sewing cotton is to form the inner insulation of the sections from one another, by being wound on before the secondary wire, and it thus fulfils the function for which rings of ebonite or gutta-percha are sometimes used.

At this point it will be well to determine the proposed contour of the finished coil—that is, of the effective part, viz., the secondary wire. The writer does not feel competent to prescribe authoritatively the ratio of diameter to length which can be guaranteed to give the best result, but one or two conclusions may be laid down which seem to be trustworthy. First, the tendency for a spark to pass from the inner edge of a section through the primary tube into the primary wire must increase with the difference of potential, which is not far from zero in the middle, and increases for a 10 in. coil to some tens or hundreds of thousands of volts at the ends. A recent writer would put the voltage of a 10-in. coil at 60,000 volts—a very low estimate.[1] Hence, although it cannot be denied that many most effective windings of secondary are sacrificed thereby, it is imperatively necessary greatly to increase the quantity of cotton wound on as the ends are approached.

Again, if a bar magnet be laid on a sheet of paper and iron filings be dusted over it, it will be seen that oval curves are traced out, roughly corresponding with the lines of force. Now it seems plain that one of these lines of force should form the external contour of the coil. For if any other arbitrary line, A B, Fig. 16, be taken as the contour, and a line of force C D be drawn intersecting

[1] See Oberbeck, *Journal de Physique*, vol. viii. (3rd series) p. 365. But this estimate is dependent on the form of the terminal, which in the circumstances of a break-down cannot of course be definite. The potential difference, however, necessary to produce a spark of a given length between given terminals has generally been put much higher. De la Rue's and Müller's experiments led them to conclude that for two balls the P. D. might be put at about 100,000 volts per inch in air, and for a point and plate 23,400 volts per inch. See Ayrton, "Practical Electricity," p. 370.

it in E F, the wire which passes outside the line of force (that is, in the two triangular spaces A E C, F B D) is producing less effect than if it were removed and rewound so as to fill the space E F.

Of course no very great exactness is possible in planning this out, nor indeed is it necessary, but it is well worth while to sketch on a drawing-board to full scale the primary tube and proposed position of ends, then, starting with, say ¼ in. for the thickness

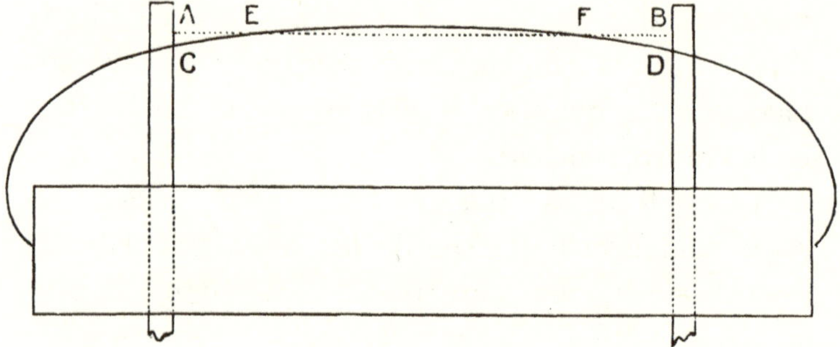

Fig. 16.—Best Shape for Secondary Coil following Line of Induction.

of cotton at the middle, to draw lines so inclined to the axis of the coi. as to reach a point, say ½ in. distant from the tube at the ends.

By laying the primary tube along the edge of the drawing-board, connecting with a battery and dusting filings on the paper, the lines of force may be sketched out, and this method the writer adopted. But the general shape of these lines is so well known, that a freehand sketch would be almost as satisfactory. Fig. 23 shows the kind of picture which will be produced, where the area shown covered with fine dots represents the best section of the secondary coil.

A word may now be said on the secondary wire. This is to be had singly or doubly covered with silk. The method of winding now being described so completely ensures that there shall never be any but the smallest difference of potential between any two portions of wire in contact, that the writer considers the single silk-covered wire preferable, as enabling more turns to be got into a given groove. It is also very considerably cheaper. The wire should, like all wire use for electrical purposes, be of the highest quality of copper. The gauge of the wire depends to some extent on what the coil is to be used for, but No. 36 is probably as good a size as can be employed for general purposes.

IX

THE WINDING

WE will now suppose everything ready to commence winding. The first thing will be to see that the trough C C′, Fig. 11, is filled with scraps of paraffin wax, of a fairly high melting-point—for the mechanical stability of the coil depends to a great extent on this—and the gas lighted. The flame should never be turned high enough to burn the wax, but only to keep it at about 110° C.

One of the centres C, Fig. 9, is now run back, and the spindle with its plates taken out of the winding machine. It is laid on a stand (which may be simply two blocks of wood), with the end e′ (Fig. 9) upwards, the plate g resting on the blocks, and e and f projecting downwards between them. The milled nut is unscrewed, and the piece k l m taken off. The plate P′ and the inner one P″ screwed to it are also removed. Only the plate P and the distance-plate Q remain on the spindle. A paper disc is now taken, one of those most strongly insulated on the inner portion, and laid on the plate P. If the hole has been correctly cut it will be found to fit nicely over the plate Q, over which it is pushed until it lies in close contact with P. The other fittings are now replaced and the spindle remounted. A reel of the dried cotton is taken from the desiccator and slipped over the spindle F, Fig. 11, and the end led through the paraffin bath C C′ under the hooks in it, passed round the edge of the plate Q on the winding-spindle, and tied with a slip-knot. A thin steel rule is applied between the plates P P′—or a marked card will answer—to plumb the depth of the groove ; this is of course done once for all and the depth known. Now, according to the position the section is to occupy in the coil, the depth of the cotton insulation to be wound on must be considered and determined on. The handle can now be turned as fast as convenient and the cotton wound on between the plates until the rule applied again shows that the layer has reached the desired thickness.

The operator sits at the end of the board with his right hand on the winch-handle, and in his left he holds a small piece of flannel folded half a dozen times into a little pad. This pad he presses on the end C where the cotton leaves the hot trough. The pad performs the double function of wiping off superfluous wax from the cotton, and putting sufficient tension on the latter to cause it to wind closely and evenly.

The writer found that in spite of the flannel pad a difficulty arose in consequence of the cotton—and subsequently the wire also—as it entered the

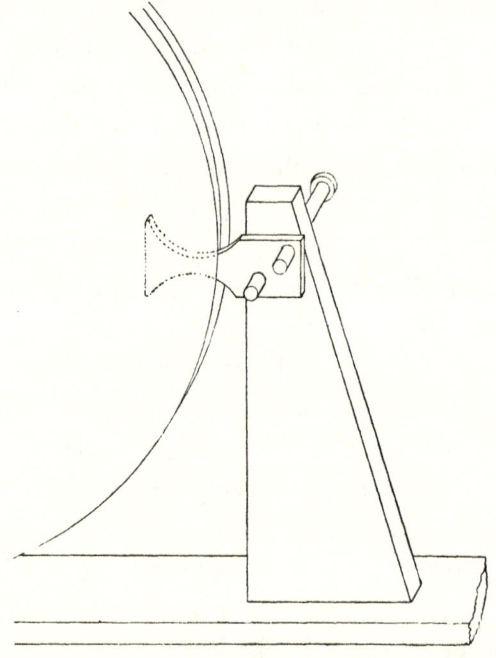

FIG. 17.—"Wiper" to keep Way Open between Papers.

groove between the plates depositing paraffin on the edges of the plates and of the paper disc. This became a source of considerable danger, the groove being so narrow, for the paraffin formed little lumps and roughnesses, which occasionally caught up the wire and produced loops and irregularities in the winding. It will be pointed out later how disastrous such loops are. But this tendency was checked by the "wiper" shown at *w*, Fig. 11, and on a larger scale in Fig. 17. It consisted simply of a piece of wood fixed upright into the baseboard of the winding apparatus near the edges of the circular brass plates A, with two brass pins passed stiffly through it near the top. These brass pins passed also through holes in a little piece of ivory (but a metal would do) of the shape shown in Fig. 17. It was about $\frac{1}{16}$ in. thick.

This little piece of ivory projected edgewise into the groove between the plates in such a position that its curved edge scraped off any deposited wax from the outer portions of plate or paper, and continuously kept the groove clear for the wire or cotton to pass into. After this was adopted no further difficulty was experienced from catching. If ever it is necessary to remove the ivory, all that need be done is, with the left hand, to pull back the two brass pins, when the ivory immediately drops off.

When the rule shows that enough cotton is wound on, this is cut and the spindle is again dismounted, and the nut, &c., removed. We then see the cotton lying as a fine close washer cemented by paraffin to the paper on which it lies. There is no difficulty in lifting off the brass plate if this has been clean and cold, for the adhesion of paraffin to it in these circumstances is slight.

The reel of cotton on F (Fig. 11) is now replaced by one of silk-covered wire, the end of which is passed through a small hole, of which there are several drilled for this purpose in the plate P′, Fig. 9, care being taken to choose one as near as possible to the outside edge of the cotton washer. The wire is threaded from the inside to the outside of the plate P′, about nine inches or so being passed through. The plate P′ is then replaced and the piece k l m put on it, the milled nut being screwed on tightly. The projecting end of the wire is given a turn or two round the waist of the piece k l m and secured in some simple way—best perhaps by being clipped under the head of one of the screws S S, Fig. 9. All is now ready to start winding. The counter is set at zero, the operator's left hand is on the flannel pad, and the right on the winch-handle. There is no reason for moderating the speed of turning. A high pace is even an advantage, for it is clearly desirable that the paraffin should have no time to cool on the wire between the trough and the spindles, or the layers will adhere less perfectly together. Moreover, practically no safety is gained by turning slowly, for the moment of inertia of the plates P P′ is sufficient, if a hitch occurs in the feeding, to break the wire even at quite a low speed ; and as the operator's eyes must be kept on the wire entering between the plates, he will almost certainly fail to notice any tangle coming until it is too late.

The winding must be continued until the eye judges that the section is approaching the desired diameter, and when the foot-rule confirms the operator's judgment that enough wire is on, the machine should be stopped and the counter read and noted in a book.

Now the spindle is again dismounted, and the inner end of the wire having been loosed from the screw S, the top plate P′ is lifted carefully off, the end of the wire slipping back through the small hole. We see, on removing the plate, the disc Q lying on the plate P surrounded by its annulus or washer of cotton, and round this is a beautifully even annulus or perforated disc of wire consolidated by paraffin fitting closely on the outer edge of the cotton washer, and extending to within half an inch, more or less, of the edge of the paper disc on which it lies.

The next operation is very important. The outer end of the wire is just as it was left from the winding; that is, nothing attaches it to the coil of which it is the extremity, but the wax through which it passed to the winder. Subsequent operations, of which there are several, are extremely likely to loosen the end, and if, in finally jointing up, the end gets pulled down towards the axis of the coil—an accident not at all unlikely to occur—the end will form a chord of the circle corresponding with the outermost layers of the section and part of it will only be separated by its silk covering from the lower layers, differing from it in potential by possibly several hundred volts. This will infallibly result in internal sparking, and burning of the silk, and will not improbably lead to a complete break-down of the coil.

The operator's attention is therefore especially directed to the avoidance of this mishap, which is perhaps the most serious that is likely to occur in his work.

To avoid this disaster he is advised to proceed as follows. A vessel should be procured or made like a small teapot, with a wooden or other non-conducting handle, in which paraffin can be melted over the gas, and some paraffin, melted to a point considerably above its fusing-point, should be poured round the section of wire so as to fill up the angle or "rabet" formed between the circumference of the wire section and the paper disc. During this operation the brass plate, with the section on it, is held horizontally in the hand. This will be found so to strengthen the section in its weakest part that there will, with ordinary care, be no danger that future manipulations will loosen the end, and lead to undesirable consequences. A very little practice will enable the operator to do this effectually without spilling any paraffin.

Next a disc is taken from the first set of paper discs, viz., one having thickened insulation on the outer portion of the surface, and laid upon the finished section, a pin-hole first having been made at a point which will come over the outer edge of the cotton washer, and the inner loose end of wire drawn up through it. The point can be obtained by measurement if the operator cannot trust his eye to determine it. The disc of paper being in proper position (the small brass disc Q, Fig, 9, coming up through its centre), it should be caused to adhere to the cotton washer and wire section. This is easily effected by ironing it with a very small laundry iron moderately heated. It is quite easy to see when adhesion has taken place, for the melting of the paraffin and exclusion of air bubbles bring the disc into optical contact with the wire, and then the colour of the copper can be seen through its silk covering and the paper disc.

The winding machine is now to be fitted together, but without the plate P″, which, together with the screws S and S′ which hold it, is removed and laid aside. The space which it occupied forms a groove just wide enough to take the next section, but another cotton washer must first be formed. The two loose ends of wire must for this purpose be carried to the left, over the plate P, and secured temporarily. The simplest way is to give them a turn or two round the spindle D (Fig. 9) and to secure the ends against the outside of the plate P by a spot of shoemaker's wax, marine glue, Chatterton's compound, or some similar substance which can be sufficiently melted by pressing the wire ends on to it by a warmed spatula.

It is highly important that this temporary fastening of the wires should be secure, and that there should be no slack, for it must be remembered that the end last dealt with comes up the groove into which the cotton is about to be wound, and if there is a possibility of the cotton catching the wire end and winding any part of it in, the wire will, in the finished coil, be brought nearer to the main ebonite tube than it would otherwise be, and the advantage of the cotton insulation will be lost. But this danger, though great, is quite easy to avoid by simply keeping the ends tight. Matters being so managed, another cotton washer identical with the previous one is wound and the cotton cut.

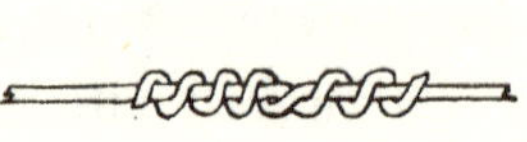

Fig. 18.—The Right Way of joining Wires.

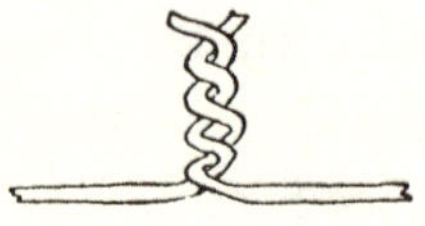

Fig. 19.—The Wrong Way.

The inner end of the first section is now loosened from plate P, to which it has been stuck, and about an inch of its end brightened by being drawn through a folded piece of the finest glass-paper held between the finger and thumb.

The cotton having been replaced on the spindle F, Fig. 11, by the reel of wire, the end of the latter is brightened in the same way, and the two ends twisted together. Fig. 18 shows the best kind of twist. That shown in Fig. 19 should never be used.

The joint must next be soldered. There are several ways of doing this, but the writer has found the following method the simplest and best. The joint being near the winding machine, it is easy to arrange that it shall be situated freely in the air within convenient reach of the hands, supported by the stiffness of the wire, no apparatus being necessary to hold the joint while it is being soldered. A Bunsen burner being alight on the table, a miniature

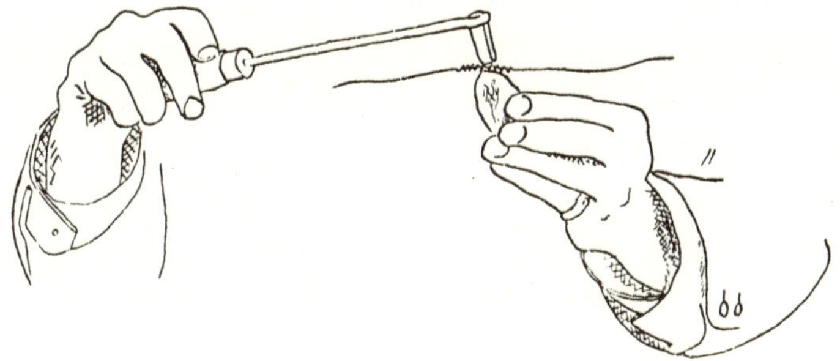

FIG. 20.—Soldering Secondary Wire.

soldering-iron well tinned is held in the right hand, and a piece of resin the size of a hazel-nut in the left. The iron is held in the flame until the colour of the latter shows it is sufficiently heated, and then the iron is brought above the joint at the same moment that the resin is brought below it. Both the iron and the resin touch the wire together, the tip of the resin melts and the liquid solder fills up the intervals between the twists of the wire on the joint. See Fig. 20. Neither here nor elsewhere in the coil should "soldering fluid" on any account be used. If it is, it is merely a question of time for the joint to break down.

The joint, if resin only has been used as a flux, requires no washing, but should be carefully covered with silk, paraffined cotton, or what is best and most convenient to apply, the narrowest and thinnest white silk ribbon, dried and paraffined. Now, if this has not already been done, the wire should be passed under the hooks in the paraffin trough and winding may be recommenced.

This section is wound exactly like the first, to the same size and with the same precautions, but with one all-important difference: *the handle of the machine is turned backwards instead of forwards.* The counter starting from zero will of course be turning backwards, but this is of no consequence, if we remember that when the section is finished the number of turns is not the reading of the dial, but the number obtained by subtracting this reading from 10,000.

The spindle is then dismounted, and paraffin run round the edge of the section from the little teapot as before, and a paper ironed on, after the loose end has been brought out through a pin-hole near the outer edge.

The writer recognises that it is quite illogical to put a paper on the top of the double section, as, when the sections are all finished and attached to the coil, the top of each double section ought to be sufficiently insulated from the bottom of the next by the bottom paper of the latter. (The words "top" and "bottom" will not be misunderstood if it is noted that, as will presently be explained, the axis of the coil is in a vertical position while the sections are being attached.) Nevertheless, there are reasons which make it desirable to have papers on both sides of each pair of sections, as, for instance, the increased safety of subsequent manipulations, and the fact that it is impossible to secure complete contact by ironing, when the sections are mounted on the tube in the positions they are finally intended to occupy.

It is desirable, though not of the first importance, that the two ends of the double section which issue the one from the groove between the first and second papers, and the other from the pin-hole in the third paper should not be quite opposite to each other, but that the line which we may suppose to join them should make an angle with the axis of the discs. Thus when the outside ends are joined up in the finished coil, they will not lie on a "generating line" of the cylinder, but on a helical line round it. This is an extra safeguard against short circuiting sparks.

The complete double section being now finished, nothing remains but to remove it gently from the bottom plate P, to put it in an envelope or other place of safety, marking on the envelope a reference number, so that the number of turns, resistance, and external and internal diameters of the wire section may be noted against it in a book.

Most of the books dealing with the subject of the Induction Coil quote from Mr. Spottiswoode's paper [1] on the giant coil constructed for him by Mr. Apps the statement that the portion of the secondary furthest from the middle were composed of thicker wire than the middle portion, and that "the object of the increased thickness towards the extremities of the coil was to provide for the accumulated charge which that portion of the wire has to carry." Mr. Apps being, doubtless, one of the greatest authorities on the practical construction of the induction coil, others may be excused—if not justified—in following what is known to be his practice. But a theoretical account of the exact function of this thicker wire would be hard to give.

There are two principal advantages in the method of winding above described over the method of winding single sections. First, the double section being symmetrical as regards its two external faces, there is no danger of the annoying mistake often made with single sections, viz., putting one or more on the tube upside down. In the case of single sections this mistake would mean that the inductive effect of the primary circuit would act upon the section so placed in the direction contrary to that of the main current, and so far from contributing to

[1] *Phil. Mag.*, January, 1877, p. 30.

the cumulative effect, it would neutralise the effect of another section, and both would be useless. The double sections may be used either way up, and their effect is the same. The second advantage is that the troublesome task of soldering the inner ends of the sections when they are *in situ*, and the risk that the joint when made will not be packed away far enough from the tube to avoid short circuiting by spark is entirely got rid of.

The first section being put away and numbered, the plates P, P′ and P″ should be warmed and rubbed with a cloth to get rid of all spots of paraffin. The winding of the second double section can then at once be commenced.

The number of sections to be made will depend upon the distance which it is proposed to include between the outside cheeks of the coil. In the case of the writer forty-eight pairs were found to fill the interval, twenty-four on each side of the central partition.

When the sections are all finished it is well to test them rigidly.

First they should be tested for resistance by the Wheatstone bridge. This will at all events make known any bad joint or want of continuity in the wire, even if the resistance itself is a matter the maker of the coil is not curious about. Nothing need be said about this, the method being described in all the text-books. But one minor point may be mentioned. No. 36 wire is too fine to fit comfortably into the binding-screws ordinarily used for connecting conductors, and the following little hint may save a good deal of waste of time in making connections. There is a little spring clip made of gilt brass and sold by most dealers in neckties, &c., and also by the vendors of studs and other small goods who line the kerbstone in the Strand, London, and other streets, which is constructed on the principle of the American clothes-peg. You pinch the tails and the mouth opens. By soldering stouter wires to one of the tails of each of a pair of these clips we have at once a simple and efficient means of connecting any apparatus such as a resistance-box and the secondary wire. The connection is made in a moment and is of a good conductivity; moreover, it does not bruise or twist the wire.

The sections should also be tested for any loops or irregularities in them which, as above mentioned, constitute a most dangerous defect. The way to do this is to put each pair singly on the primary tube,—in the middle, for there the induction is greatest—and then work the break of the coil to its fullest power in a darkened room. This should be done both with the ends of the section-wire brought within sparking distance, and with the ends separated too far for a spark to pass. If any internal sparking is seen through the end papers the section should at once be condemned and rewound.

Should any rewinding be necessary, the task is not a formidable one. All that it is necessary to do is to put the section back into the winder, lead the end out and attach it to an empty reel on F (Fig. 11), and turn the handle G of the "unwinder." Of course the wire does not pass through the bath on this occasion. The wire need not be wasted, but can be at once rewound between fresh papers, using the same precautions as at first, and, of course, passing it through the hot bath.

X

PREPARING THE PRIMARY TUBE FOR MOUNTING THE SECONDARY.

TO mount the sections on the tube it is necessary that the latter should be in a vertical position. For this purpose a stand must be made. A simple way is to cut two pieces of wood eight or nine inches square, and through the centre of one of them bore a hole just big enough to take the tube without shake. If there is to be no central partition this may be a plain hole, but if a central partition is to be used, saw-cuts should be made from the hole to one side forming parallel tangents to the circular hole, so that the tube can, if desired, be removed sideways. In any case this facilitates the manipulation, for as the sections are attached the weight increases to an amount which makes the work unpleasantly heavy to move, and a drop on the floor would be fatal. Two other pieces of board should be used to connect the two pieces already mentioned, as shown in Fig. 21, so as to form a kind of stool with a floor as well as a top.

The distance between the upper surfaces of the two horizontal boards should be about three-quarters of an inch more than half the length of the ebonite tube, this three-quarters of an inch being occupied by the two wedges shown on the

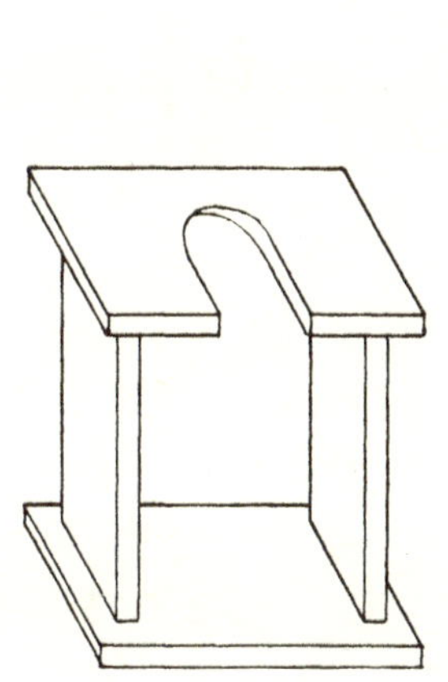

FIG. 21.—Stand for Supporting Coil in Vertical Position.

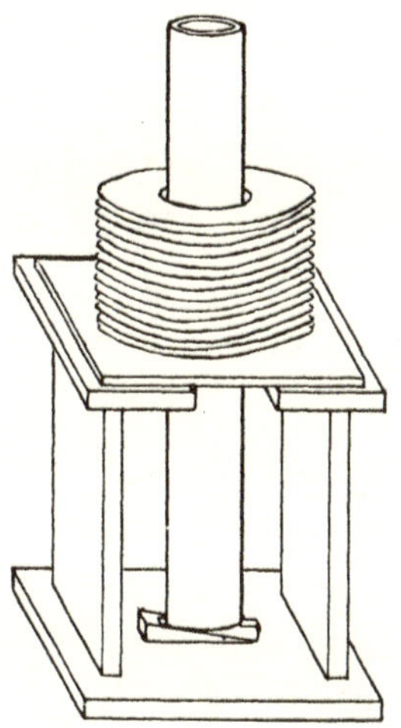

FIG. 22.—Coil on its Vertical Stand.

base in the Fig. 22, by which the distance can be rendered exact.

The tube is now placed on the wedges and fastened by string to the stool. A square is applied so as to make sure that the tube is quite perpendicular to the plane of the top board. If a central partition is desired it will have been cut exactly to shape, with a hole, made in the lathe, of such size as to fit stiffly over the tube. One object of the central partition being to assist in supporting the weight of the coil and to give extra assurance against deflection, with the consequent possible cracking of insulation, it is of course highly important that all three plates—that is to say, the central partition and the two end-cheeks, to be presently described—should have an equal bearing on the baseboard. This involves the necessity that their lower edges should all be exactly the same distance from the axis of the holes to be cut in them to admit the primary tube. One way to secure this is to decide upon the exact height which the axis of the coil is to be above the baseboard, and carefully record it. This height will govern also the height of the axis of the break. This height being once determined on, the three plates should be got out to size at one time, and the centre of the holes which it is intended to cut scribed very carefully with a gauge, or scribing-block and surface-plate, from the lower edges. But a more accurate method is as follows :—

We will suppose the height in question is fixed at six inches, which is suitable for a coil of the dimensions of the one being described. The central partition is taken, and six inches carefully gauged from its lower edge by a line drawn parallel thereto. A vertical line is then gauged by trial equally distant from each vertical edge.

The intercession of these two lines should now

be marked with a centre-punch, and will determine the point which will be the centre of the holes to be cut.

Now mount on the lathe a wooden face-plate turned true, and attach the central partition to it by two or three little slips of wood screwed over it into the face-plate. The partition must now be very exactly adjusted by light taps with a hammer until the centre-punch mark is precisely in the centre. The lathe is then stopped, and the face-plate having been removed, two slips of wood are screwed to it so as to lie closely along two adjacent edges of the partition, one of these being the lower edge, and the other one of the vertical edges. Then the face-plate is returned to the lathe and the hole cut in the partition. The latter can then be removed and the face-plate laid aside. When the time comes to cut the necessary holes in the end-cheeks, all that will be necessary will be to lay them, one at a time, on the face-plate, allowing their edges to rest against the slips of wood left on the face-plate for this purpose, and fasten them down in exactly the same way in which the central partition was fixed. There will thus be no doubt that the three holes will be exactly co-axial. Seasoned wood should be chosen for the face-plate, which will not get out of shape during the time elapsing between the two operations.

Before the central partition is fixed to the ebonite tube, any binding-screws, &c., which are desired to be attached to it should have been polished, lacquered, and fixed. The partition can be attached to the ebonite tube (more or less) with marine glue or indiarubber solution, and, the tube being fixed vertically in the stand, the partition can be adjusted until it is quite central and at right angles to the axis of the tube. The best possible polish will of course have been given to all parts of the partition which will show in the finished coil, but the part against which the inner section has to lie should be roughened, so as to increase the adhesion of the paraffin.

It may here be mentioned that ebonite should be treated after sawing and filing to shape, first with glass-paper, ending with the finest procurable, then with Bath-brick and water, and finally with rotten-stone and kerosene. With skill and this treatment the finest black polish may be obtained.

XI

MOUNTING THE DISCS

THE partition being thus attached (or, if it is not desired to have one, a temporary partition of tin plate), one end of the first double section is attached by solder to one binding-screw of the partition, or otherwise left loosely projecting, and the double section laid on the partition centrally. A little hot paraffin may be poured on the partition just before the section is laid in place, which will help to stick the latter down.

The central apertures in the sections being rather bigger than the tube, melted paraffin can be run in from the little teapot into the annular space thus left. The paraffin should be heated as highly as possible without decomposing it, that it may run not only into the annular space, but between the first double section and the ebonite partition. If no partition is used this is less important. The other sections can now be treated exactly like the first, a pin-hole being made in that one of the extreme papers of each section which at present has none, and the end of the wire projecting from the pin-hole in the next section passed through it. Thus the two wire-ends which are subsequently to be soldered will lie closely together *in* a section and not between the outer papers of two consecutive sections. This is better than leaving them between the sections, as there is thus no chance of the twisted end getting nearer the core than it should be. As each section is put in place, hot paraffin is run in from the little teapot to fill the interval between its inner edge and the ebonite tube. The paraffin will also run in between the sections and consolidate them. When the first set are all attached the tube is dismounted and reversed, and the other half of the sections treated in the same way. This being done, all the projecting ends of wire should be twisted and soldered together in pairs, and the coil lifted off its vertical stand and replaced on its Y's. The secondary can now be tested for continuity and resistance, but will probably not stand sparking, the insulation between the outsides of consecutive sections being insufficient. To improve this, dry cotton yarn should be wound round the grooves left between the paper discs, being drawn on its way through a paraffin bath, as described in the next section. But first, to avoid accidents to the very fine wire forming the two extremities of the secondary coil, it is well to take a yard or so of thicker wire, say No. 22, and having soldered one end to an extremity of the secondary, take it once or twice round the groove between the last paper and last paper but one, bringing the end, a foot or so long, out through a hole near the edge of the last paper. Each end of the secondary should be treated in this way.

XII

THE OUTER INSULATION

NOW the coil should be mounted horizontally with the projecting ends of the tube resting in some supports so contrived that the coil may be revolved as if in a lathe. To avoid spoiling the ends, it is well to construct a rough square framework of deal with bearings carrying anti-friction wheels consisting of cast-iron "sheaves" in the grooves of which thick rubber rings are laid, just as the rubber tyre lies in the groove of the wheel of a solid-tyred bicycle. Both the wheels and rubber rings are easily procurable.

The coil being so mounted, a reel of loose cotton (sold for covering wire), carefully dried, is placed in a paraffin bath, kept hot, and the end being inserted in one of the grooves between the paper discs the coil is pulled round until the groove is filled with cotton, nearly, but not quite up to the edges of the discs.

The projecting solder joints will be wound in and covered by this cotton.

Some softening of the paper discs is inevitable in this part of the process, owing to the heat brought in by the hot wax on the cotton; but this is not of importance, as when a disc becomes soft it will adhere to its next neighbour, and just as good insulation will result. Finally, a strip of macintosh two or three inches wide and long enough to go round the coil, should be taken and held tightly round a portion of the coil so as to embrace closely the edges of the paper discs, except just at the top, where a gap is left big enough to permit of hot paraffin being poured in, the macintosh serving the same purpose as the clay roll used to confine the melted lead with which lengths of water main are joined.

When this is done, and the paraffin is cool, the macintosh is moved along the coil and the process repeated. In this way the whole coil becomes covered with a solid insulation of paraffin half or three-quarters of an inch thick. A search should be made for any crevices which may have been left, and hot paraffin from the little teapot poured into them after enlarging them, if necessary, with an ivory paper-knife.

The coil should then be brought to a truly cylindrical shape by turning it in its stand, while a broad chisel is held against it so as to dress off projecting portions of paraffin. This being done, and the operator being satisfied by inspection that no visible cracks remain in the paraffin covering, the coil may be tested by putting it on its temporary stand and connecting it with a full battery.

XIII

COVERING AND FINISHING

SO far as practical work is concerned the coil is now finished. But it should, of course, be mounted and covered, and we proceed briefly to indicate the steps which may be taken for this purpose, premising that the final shape and design of the exterior are largely a matter of taste. One method will therefore be shortly described, but without implying that other methods may not be just as good.

Two[1] strips of sheet ebonite are taken long enough to go almost round the coil, and about 6½ in. wide; that is to say, a sixteenth of an inch wider than the distance from the central partition to the outside of the last paper. Small holes are drilled half an inch apart along the short edges to serve for lacing, and, the coil being again mounted in a vertical position, one of the sheets is softened in hot water and, having been quickly rubbed dry with a

[1] That is, if a central partition is adopted. If it is not desired to have one, a single sheet wide enough to cover the whole coil with $\frac{1}{16}$ inch to spare at each end will of course be used.

hot towel, is wrapped round the upper half of the coil. It is there held until cool, and then tightly laced with fine catgut ("A" string of the violin). The lacing is of course in that part of the circumference which, when the coil is finally mounted on its stand, will be underneath. The coil is then turned over and the other half treated in the same way. Thus the coil is now covered with two cylindrical ebonite covers, fitting closely against the sides of the central partition, and projecting a sixteenth of an inch at each end beyond the last paper disc.

Now when the central partition was made, two other pieces of ebonite will have been prepared of the same dimensions exactly, and having a mark to show the centre of the hole to be bored in them; the only difference being that whereas the central partition was cut from $\frac{3}{16}$ inch sheet, the two others are from half-inch sheet. Only considerations of expense limit the thickness to half an inch; the coil would have more mechanical strength if these ends were three-quarters or even one inch thick.

Two pieces of ebonite tube are now procured $3\frac{1}{4}$ in. in diameter and 4 in. long, being so chosen that they will slip easily over the projecting ends of the main tube of the coil. These short pieces are destined to be screwed into the end-cheeks and to cover the projecting ends of the inner ebonite wrapping of the primary coil, as presently described. They add greatly both to the mechanical and electrical strength of the coil, besides much improving its appearance. For when the main tube, with the primary and core in it, is treated with hot paraffin in the manner described on p. 27 above, it is not easy to preserve its circularity of section. And even if this is preserved, it is most difficult to form a joint between the main tube and the end-cheeks which shall be mechanically and electrically strong, while at the same time the finished effect of the truly circular end-tubes is pleasing to the eye. Hence the small end-tubes recommended are almost a necessity. But if these should be unprocurable, a makeshift may be constructed by wrapping softened ebonite sheet round a mandril nicely turned and of just the right size. This sheet should be at least $\frac{1}{8}$ in. thick, and cut so that the joint "buts" exactly. It will then be possible to thread one end of each to screw into the end-cheeks, and to attach to the other ends the finishing caps described presently. The threads should be cut on these pieces, if one is driven to use them, while they are still tightly bound on the mandrill. Much care would, however, be required in screwing in the pillars presently mentioned which cover the ends of the primary wire if the tubes had to be made of sheet, and solid tubes should certainly be used if it is possible to obtain them. The thick plates, or end-cheeks, are now chucked on a face-plate in the lathe—on the wooden one described above (p. 91) if the method is adopted which is there recommended for securing that the holes are co-axial—and the holes cut and threaded in them, similar threads being cut on one end of each of the short tubes, which are then screwed into them. A hole $\frac{5}{8}$ in. in diameter is then drilled and tapped in the under side of each of the short tubes to take a little turned pillar of ebonite, down which a hole has been drilled large enough to convey the three free ends of the primary at each end of the coil. These pillars have a shoulder turned on them in such a position that when screwed into their holes the shoulder shall rest on the baseboard when the edges of the two end-pieces and the central partition do so—or, so that the bottom edges of the central partition, the two ends, and the shoulders of the two pillars shall all be in one plane. The pillars are prolonged beyond the shoulders so as to pass through holes in the baseboard, and are threaded to receive ebonite nuts underneath.

The coil being mounted on its vertical stand, one of the end-pieces, consisting of the cheek with the short tube screwed in, but without the pillar, is slipped over the projecting main tube, a cork being inserted in the hole made to take the pillar, and the piece is adjusted until the bottom edge is co-planar with that of the central partition. The end of the secondary is brought to the top of it by being threaded through a hole made for the purpose as shown in Fig. 23.

Melted paraffin is now poured in at the top, so as to run into the space between the last paper disc

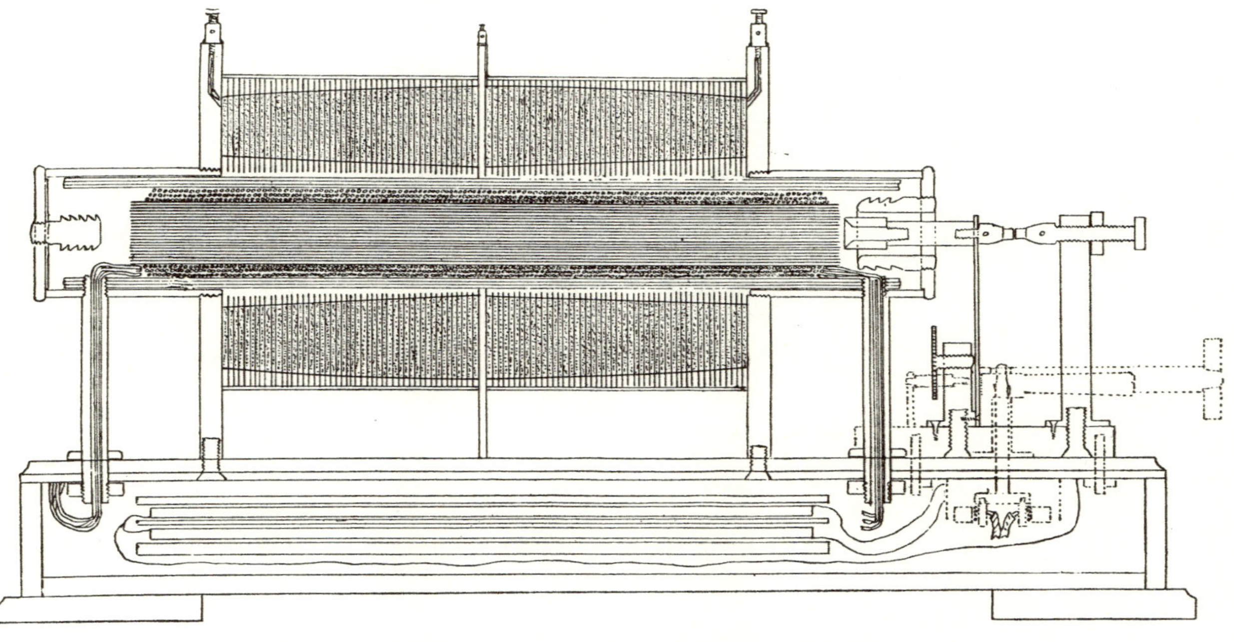

FIG. 23.—Section of Complete Coil (terminals are omitted for clearness).

and the ebonite end-cheek, also between the short tube and the main tube, thus holding all firmly together. This is repeated at the other end of the coil, special pains being taken to secure that the bottoms of both end-cheeks and the central partition shall be co-planar. This may be secured by tying a drawing-board against the bottom of the central partition and the end-cheek already fixed, and adjusting the remaining end-cheek against it; or the baseboard itself may be attached before wax is poured into the end-pieces. See p. 107.

Holes are now bored through the main tube and tapped to form continuations of those tapped to receive the pillars, and the three loose ends of the primary at each end of the coil are brought through them, sufficient paraffin inside the end of the main tube being melted out to admit of this.

Ornamental ends can now be turned out of $\frac{3}{16}$ sheet ebonite and shouldered to fit into the short end-tubes. One is drilled and tapped to take a $\frac{3}{8}$ Whitworth thread; the other for the break end must have a much larger hole cut, large enough to screw over a short piece of ebonite tube $\frac{13}{16}$ in. in diameter inside and $1\frac{5}{8}$ in. outside (Fig. 23).

These ends are to be fitted in the following manner: The small piece of tube last described having had deep corrugations turned or filed in it, except at the end where it was threaded to fit the end disc, is screwed into the disc, but not quite home. The coil being on its vertical stand, the paraffin at the break end is melted by the insertion of a hot iron and the tube with the disc screwed to it placed in position, the shoulder on the disc securing that the disc shall be central, and the disc holding the tube central. When the paraffin sets the tube is held firmly by its corrugations, and the disc can be screwed home. A cork will have been put into the little tube to prevent the wax filling it up, and this cork is afterwards drawn out, so as to leave an opening for the hammer-head of the break. A small chisel or other tool should be worked about in the hole where the hammer of the break is to work, so as to remove any paraffin, or ends of tape, which may be covering up the ends of the core wires; for it is desirable that the hammer-head, when the coil is in action, should be about $\frac{1}{16}$ in. from the core, nothing intervening which can in any way impede its freedom of movement. The disc at the other end is fixed in the same way, except that instead of a tube a short solid rod of ebonite is used.

It will be seen that this method has the advantage that great obstacles are set in the way of a spark passing between the terminals of the secondary by way of the iron core. At the break end (when the mercury break is used, as it would be for the longest sparks) the spark must pass from the terminal at the top of the cheek to the opening in the end of the little tube, and then along the inside of this little tube to the end of the core, a total distance of

9 inches ; while at the other end there is solid insulation of paraffin and ebonite unweakened by any metal work whatever, and quite impassable by any spark the coil is capable of producing.

The base of the coil is now finished and polished, two holes being bored to take the bottoms of the two pillars, and all the fittings—commutator, binding-

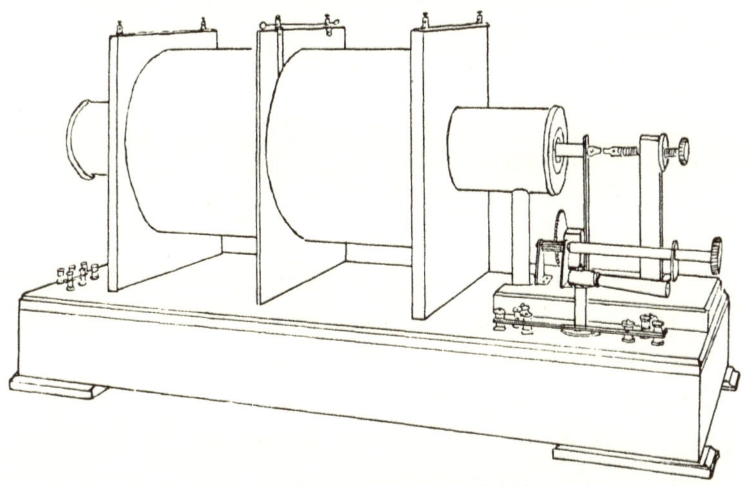

FIG. 24.—Complete Coil, perspective view.

screws, &c.—attached in their right places. The coil is then placed on the base if not already attached to it, the two ebonite nuts screwed on to the pillars, and $\frac{3}{16}$ Whitworth screws inserted through the baseboard into holes drilled and tapped for this purpose in the bottom edges of the end-cheeks. A perspective view of a coil finished according to the method above described is shown in Fig. 24.

The writer has seen it suggested that polished wooden uprights should be used set just outside the end-cheeks to support the primary tube, and by it the whole coil. It must be conceded that this method has great advantages hoth in point of cheapness and because it secures great mechanical strength. But wood, even the driest, is so vastly inferior as an insulator to ebonite, that the balance of opinion is decidedly unfavourable to this method of mounting, which, so far as the writer is aware, is never adopted in the best large coils. It is, no doubt, quite suitable for small ones.

A coil made in exactly the way described has proved capable of standing a fair amount of rough usage, and no accident has occurred to it although it has been driven in a cart several miles. Nevertheless, it must be admitted that considering the great weight of a large coil, it is and must be an apparatus which is structurally weak. It is particularly ill-adapted to resist such shocks as would arise from an impact of the end of the base against any resisting object—as, for instance, if it were pushed along a table and the base came in contact with a heavy piece of furniture; or if while it was being carried about, the end of the base were to strike against anything. The weight of the bobbin is very considerable, and although it resists deformation sideways with fair sturdiness, this is not the case with end-long stresses. One would be so much better pleased if one were at liberty to strengthen it with some stout angle-irons screwed against the face of the end-cheeks, but the paramount importance of insulation forbids this. All that can be done is to have the end-cheeks as thick as possible, and to shield the instrument when in use with the greatest care from all mechanical strain. It would no doubt be possible if the coil is destined for rough work to add a little strength by screwing

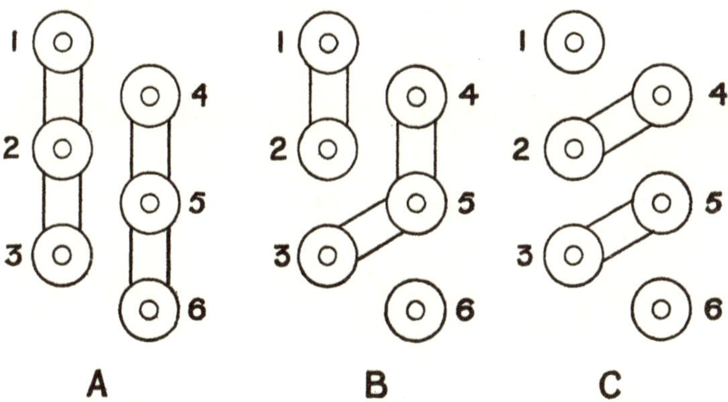

FIG. 25.—Terminals arranged—A, for the three layers of primary used as one; B, used as two; C, used all in series.

to either the inner or outer face of the end-cheeks angle-brackets of stout ebonite sheet, say $\frac{3}{4}$ in. thick, and fastening them to the baseboard by screws from below. But a great deal depends upon the truth with which the lower edges of the end-cheeks have been cut, and the even bearing which they take on the surface of the baseboard.

Nothing remains to be done now but to make the connections under the base. These will present no difficulty to any one who understands the theory of the coil, which is the same for a large as for a small one. It need only be said that if three separate layers of primary wire are used, their ends should be carried to six terminals close together at one end of the base arranged as in Fig. 25, under the heads of which connecting pieces can be readily clamped, these pieces being of uniform length whatever arrangement of primary is desired. On comparing the three diagrams, Fig. 25, with that of the connections under the base shown in Fig. 26, it will be seen that the course of the current from the battery is as follows: A B, Fig. 26, being the battery terminals, we will suppose that the current enters at A. It first passes to the commutator C, which we will suppose is so set that the pin connected with the wire from A is pressing on the spring which is to the right hand in Fig. 26. Thence it passes to F, which is one of the terminals, which serve also as clamps to hold down the base of the break. From F it passes up the pillar of the break (if an Apps) and by way of the platinum studs and the hammer-spring to E. E, it will be noticed, is connected with an end of one of the layers of primary wire coming down the ebonite pillar seen projected into the small circle at P; it is also connected with No. 6 of the family of six terminals shown also in Fig. 25. Let us suppose first that these terminals are connected as shown in Fig. 25, A. Nos. 4, 5, 6 being in metallic contact, the current divides

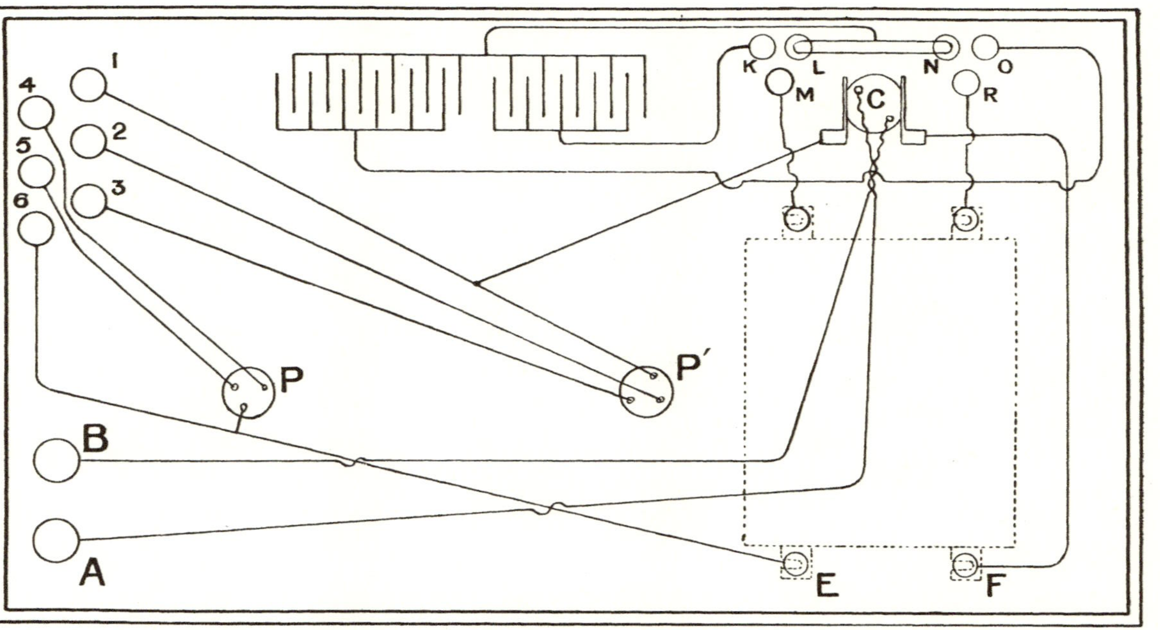

Fig. 26.—Diagram of Connections under Base. A B, battery terminals; C, commutator; E F, screws holding down base of break, and making connection therewith; K L M N O R, condenser terminals; 1, 2, 3, 4, 5, 6, terminals allowing different arrrangements of primary; P P′, wires coming down from primary coil through ebonite pillars.

between them, and passes through all the three layers of primary as if they were one large wire emerging at P′, whence, since Nos. 1, 2, 3 terminals are also connected, it reunites and passes by way of the other pin of the commutator and battery terminal B to the battery again.

Now suppose the terminals Fig. 25 are connected as at B. The current then reaches screw No. 6 as before, but cannot leave it. It therefore passes by the connection shown at P along one layer only of the primary to P′, and so to No. 3. This is seen to be connected with both 4 and 5, so that here the current divides again, passing through the other two layers of primary to P′, and so (Nos. 1 and 2 being joined) to the commutator and battery as before.

The arrangement shown in Fig. 25 C need not be described in detail, but it will be seen that here the current has to pass along the three layers of primary in series.

The means is therefore provided by simply altering the little connecting pieces shown in Fig. 25, of using either one, two, or three layers of primary, and any combination of them in parallel or in series. The difference in result is very marked, mainly on account of the great changes produced in the self-induction of the primary coil. See Appendix I., p. 145. It is, on the whole, better (though the opposite plan is more usual) to put the battery terminals at the ends of the base furthest from the break, so that the wires are not in the way while working. In connecting up, two things should be borne in mind—first to keep all brass work as far from the secondary terminals as the size and shape of the baseboard will allow ; and secondly, to keep the resistance of all primary connections as low as possible by using everywhere large wire, or ribbon, of good conductivity and making every solder-joint perfect.

It need hardly be added that no zinc chloride or other soldering fluid should ever come near any part of a coil, but that resin only should be used in soldering.

Fig. 26 also shows the connection of the condenser. The condenser itself is shown in two parts —of course diagrammatically. It will be seen that the terminals L and N being permanently connected under the base and to a sheet near the middle of the condenser, if K is connected with M and N with R, the right-hand half only of the condenser will be used. But if the connecting strips be shifted so as to connect L to M and O to R, then the left half only will be in use. To bring the two halves into "cascade," K may be joined to M and O to R. To use the full capacity of the condenser, L is to be connected with M and O with R, and by a longer strip K is to be joined to O.

A moderate-sized coil may, of course, be turned carefully over sideways while the connections under the base are being made, but with a large one this can only be done with some risk of mechanical strain, and it is safer, though rather more troublesome, to adopt the method employed by the writer of slinging the bobbin by webbing like the girths of a horse's saddle from above, either suspending the girths from a pole laid across two tables, or, more conveniently if practicable, from a cross-bar attached to light tackle hung from a stout hook in the ceiling.

With regard to the battery power required to work a coil of large size, it must be remembered that for many experiments the full power of a large coil is not needed, and it will generally perform satisfactorily with half the battery power required to bring out its full capabilities. The coil above described may be worked with as few as two Grove cells, with which it gives an $8\frac{3}{4}$ inch spark between point and plate, but for the best effect five cells are needed. In any case it is clearly desirable to use large cells, as the primary resistance is so low that if small ones are used the resistance of the battery becomes of importance.

XIV

THE COST OF CONSTRUCTION

A FEW words may be said about the cost of construction. The estimate of course applies to a coil of the size which has been specially treated of in this handbook; it is left to the reader to form his own estimate of what the cost is likely to be if he departs from the dimensions specified. The prices given are those paid in London. It is usually considered that for coils of moderate size, one mile of secondary wire should be allowed for each inch of spark-length desired. But it must be remembered that this is only approximately true, and it will be found that the number of miles in a large coil will increase more rapidly than the inches of spark-length. And that this must be so is obvious, if it is remembered that the inductive effect depends, not upon the length of secondary wire, but upon the number of turns it makes round the core, and the mean strength of the field in which those turns are situated. Now in both these respects it is clear that the advantage is with the small coil. As a diameter of a turn of wire increases, so does its length, and so does its distance from the region where the magnetic field is strongest. In the great coil made for Mr. Spottiswoode in 1876 the exterior turns took, each of them, more than five feet of wire; so it is not difficult to understand that to obtain 341,850 turns, no less than 280 miles of wire had to be used, and the maximum spark-length attained was 42 inches.[1]

The following, then, may be considered a safe estimate of the expense likely to be incurred:—

	£	s.	d.
The core, straightened and bound up	0	5	8
6 lbs. No. 14 best single silk-covered wire (primary)	0	15	0
Two sheets of ebonite No. 16, B.W.G. (4s. per lb).	0	11	6
¼ ream bank post paper for condenser	0	2	0
9 lbs. tinfoil	0	14	0
Platinum for break 1 in. long, × ¼ in. diameter (⅓ oz.)	1	8	4
Paraffin (say 10 ibs.—there is always waste of this material)...	0	6	8
17 lbs. No. 36 single silk-covered wire (7s. per lb.)	5	19	0
Ebonite for the ends, pillars, &c., about... ...	1	4	0
Mahogany, say	0	5	0
	£11	11	2

To which must be added a sum necessary to cover small incidental expenses for brass rod, castings,

[1] *Phil. Mag.*, January, 1877, p. 30.

terminals, &c. Probably about £12 would be spent. The winding machine is not here included, as it will be quite as serviceable, if well made, after as before its use on a single coil.

XV

MECHANICAL BREAKS

OF vibrating breaks the Apps model (with or without the slight modification suggested above) fulfils all ordinary requirements. But of mercury breaks there are many designs. One of the simplest and oldest is merely an Apps break, with the pillar and contact-screw removed and a piece of stiff wire, bent in the shape of a quadrant, attached by one extremity to the back of the spring behind the hammer-head, the other extremity passing down into a cup containing some mercury covered with a non-conducting liquid such as alcohol. The splashing of the alcohol (which should be avoided on account of the injury it inflicts on the French polish of the stand) can be prevented by a disc of cork or other material fixed on the wire a little above the level of the liquid, and by attaching to the top of the cup a cover perforated to serve as a "baffle plate," while allowing the wire free motion. Iron is a good material for the cup, and an ordinary "cap" as sold by dealers in gaspipe, that is a socket closed at one end, will answer very well. The rapidity of oscillation can be to a certain extent regulated by loading the hammer-head.

It may be observed, however, that no mercury break can be considered satisfactory which is not provided with some means of closing the cup when not in use to prevent the evaporation of the alcohol—for which, by the way, kerosene may be substituted with advantage.

There is a method now becoming common of designing coils with an Apps spring break at one end of the core, and a break identical in principle with that just described at the other. On this it may be remarked that it is unquestionably better to make the break removable than to open up both ends of the core, thus providing a tempting path for errant secondary discharges.

Perhaps the best-known mercury break is the Ruhmkorff-Foucault form, described in the *Comptes Rendus*, vol. xliii., p. 44, of which a figure is given opposite.

The rate of oscillation is adjusted by the little weight seen at the top of the figure, and the distance to which the points dip into the mercury (an important adjustment) can be regulated by the rack and pinion seen in the centre. Foucault and Ruhmkorff used platinum points to dip into the

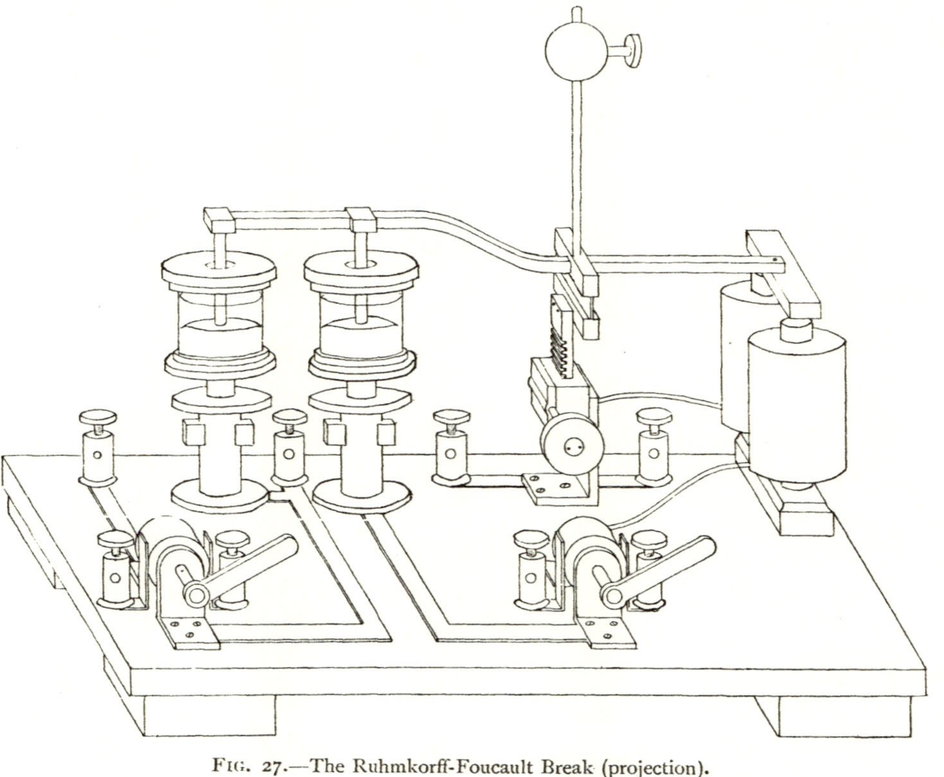

FIG. 27.—The Ruhmkorff-Foucault Break (projection).

mercury, but this is an unnecessary extravagance, as it will be found that silver answers every purpose, the local heating being less than in the case of the spring break.

The great objection to this, as also to several other well-known forms of mercury break, is that an independent battery must be used to drive it. The current which is actuating the coil cannot, without loss, be sent through the coil of an independent magnet, partly because of the necessarily increased resistance, but mainly because the increased self-induction interferes with the suddenness of the break on which the inductive action of a coil depends.

The same objection holds against an otherwise excellent break recently introduced by a well-known Paris firm of instrument makers, and described in the *Journal de Physique*, 3rd series, vol. viii., p. 336 (June, 1898), where a small motor driven by an independent battery communicates, by means of an excentric pin and slot, a rectilinear reciprocating motion to a vertical shaft dipping into mercury covered with an insulating liquid.

To avoid the trouble of a second battery, a break was introduced many years ago in which a vertical plunger was driven by clockwork and the speed regulated by a fan. This may be seen figured in Gordon's "Electricity and Magnetism," vol. ii., p. 46, but as the method of constructing it would be sufficiently obvious to any one desirous of making one, it is not considered necessary to reproduce the figure here.

The writer offers as a suggestion the following

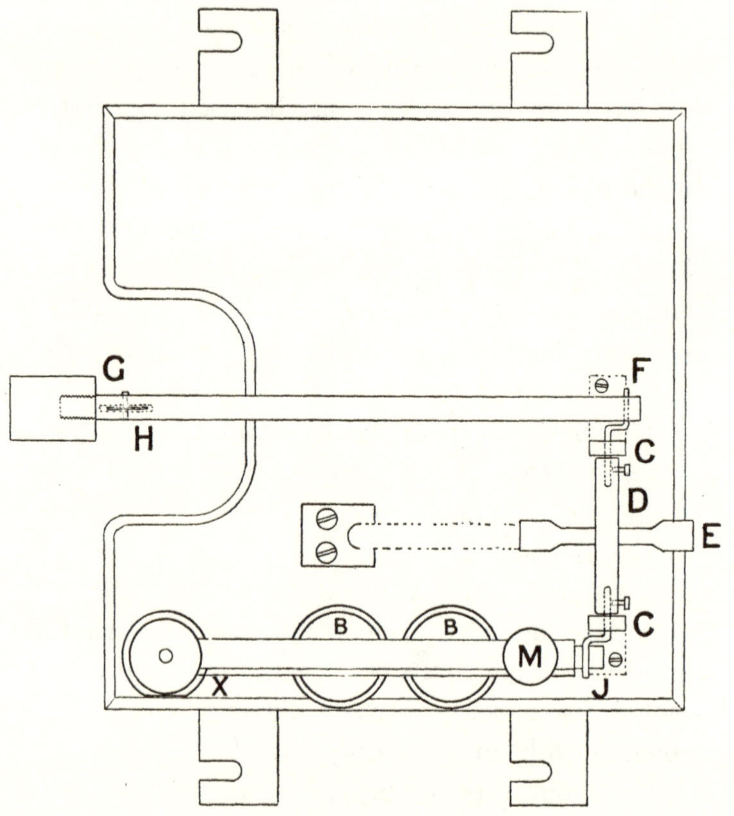

Fig. 28.—Plan of Self-acting Mercury Break.

design, which works well, and requires neither clock nor second battery. Moreover, it is designed to produce the most sudden break of contact possible. Fig. 28 is a plan of the apparatus;

Fig. 29 is partly an elevation, and partly a vertical section through the line G F, and Fig. 30 a vertical section through the two mercury cups. The same letters apply to the same parts in all three figures. It is practically a "horizontal engine" worked by the

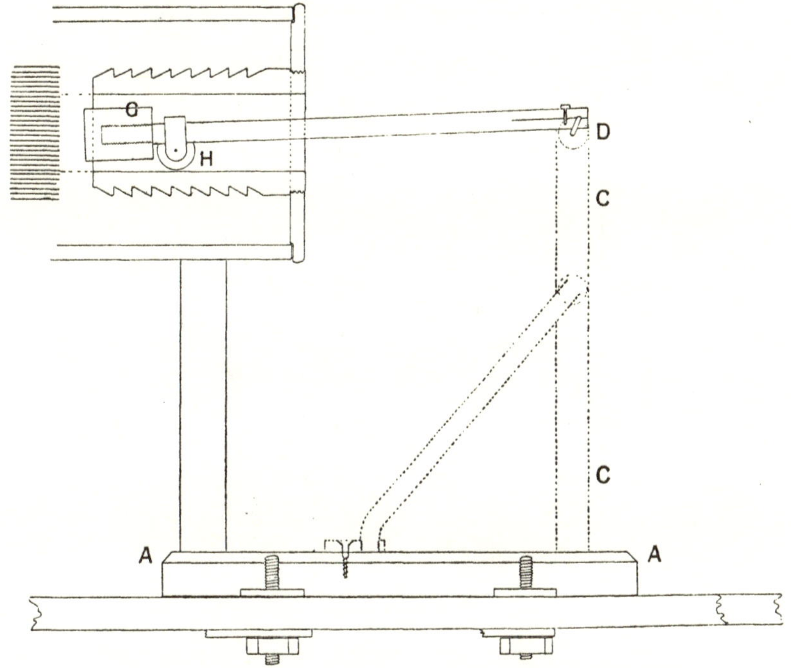

FIG 29.—Self-acting Mercury Break; vertical section through plunger.

magnetism of the core itself, and breaking contact by allowing a spring to jerk a stirrup suddenly out of two mercury cups which it has been bridging.

A A is a baseboard similar to those of the other breaks carrying the slotted brass clips for clamping it to the coil (see Fig. 6, p. 44); the brass strips are in communication by means of two binding-screws and bent wires W W, with mercury contained in two cylindrical vessels B B. The

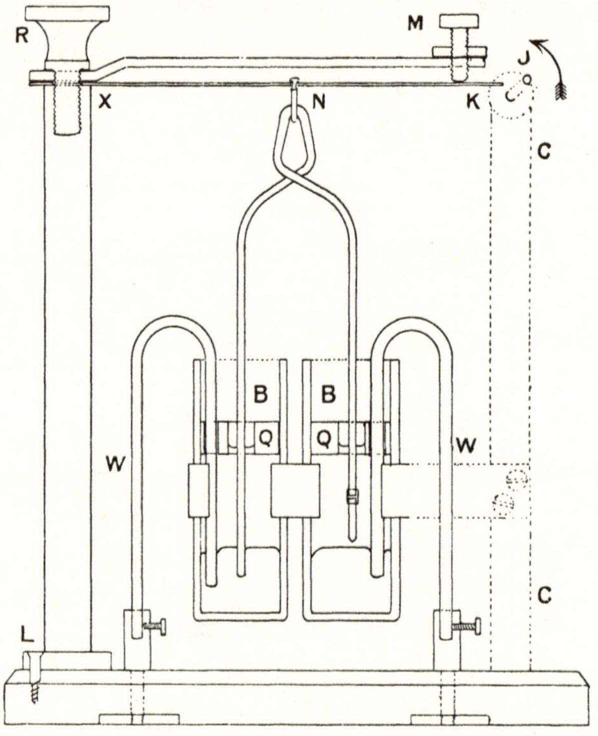

Fig. 30.—Self-acting Mercury Break; vertical section through mercury cups.

mercury is covered with kerosene oil. Two standards C, C′ (seen projected into one in the elevation at C) form bearings for a little shaft D, on which is fastened a flywheel E. At each extremity of the shaft is a little wire crank. The front one only can be seen in the elevation at D. The back one seen in the plan at F passes through a hole near the end of a light ebonite rod F G, carrying a small piece of soft iron similar to the hammer-head of a spring break at G. A little ivory castor or roller H (a small card-counter) is attached by a collar of ebonite to the rod F G, as shown. The other crank J has nothing attached permanently to it, but as it revolves in the direction shown by the arrow, in one part of its revolution it engages the top of a piece of clock-spring X K, clamped at the end X to the top of the pillar X L, pressing it down and bending it until, when nearing its lowest position, it escapes from the end of the spring, which flies up, quite free from engagement with any part of the mechanism, until it strikes against the point of the adjusting-screw M. A little ring attached to the spring at N has a piece of bent wire (aluminium for lightness) hanging from it, one end being rather longer than the other. The shorter end is tipped with silver, and the level of the mercury in the cups is so arranged that the longer end of the aluminium wire is always immersed, while the shorter silver-tipped end is just withdrawn when the spring is nearing the end of its upward travel and is consequently moving very fast. Thus the break is extremely sudden. The "lead" or "phase" of the two little cranks is so adjusted, that the period during which contact is made coincides approximately with that during which the piece of iron G (which projects into the end of the tube and is close to the core wires when furthest advanced into the coil) is approaching the core, the attraction of which on it drives the whole machine. At the moment when the piece G is almost in contact with the end of the core wires the spring escapes, contact is broken, the core becomes demagnetised, and he inertia of the flywheel carries the cranks round the rest of the way.

There are two points to note in managing a break of this kind. First, never to work it with the condenser disconnected from the coil. If this is done, the spark at the surface of the mercury is so great that an explosion will probably take place in the cup—due no doubt to the sudden volatilisation of the oil—and the cup, if of glass, will be broken, and if of metal, the cork baffle will probably be blown out and the liquid scattered about the room. Moreover, even when the condenser is attached a spark will occasionally crack the cup. In fact, although glass has often been recommended for these mercury cups, the writer has had so many accidents with it that he strongly advises that iron cups should always be used, notwithstanding that glass has a much neater appearance, and it is tempting to employ glass on account of the accurate fit of the stopper. The second point is that it is well to allow a small iron washer to float on the surface of the mercury in the cup where the break takes place, the silver-tipped wire passing through the hole in the washer. This is to prevent waves forming on the surface of the mercury. These waves, if allowed to form, will render the level of the mercury uncertain and produce contacts while the piece G is receding from the core, thus stopping the machine. The washer should be of outside diameter somewhat less than the inside diameter of the cup, and have a central hole about $\frac{1}{4}$ in. or $\frac{3}{8}$ in. in diameter.

While the piece G is moving backwards and forwards to and from the core, its weight is supported on the little castor H which runs along the bottom of the small tube mentioned on page 105.

The pace of the break can be regulated by fixing it on the base of the coil nearer to or farther from the core by the studs and slots which attach it thereto, and when in good order the speed of the break is very uniform, allowing a complete saturation of the core with magnetism and breaking contact more suddenly than is possible with a hammer break. With the hammer break there is always a slight want of suddenness in the break of contact due to the formation of an incipient arc, composed of gases at a high temperature which conduct the current after metallic contact is broken. Perforated corks, Q Q, should be inserted in the two cups, with holes about $\frac{3}{16}$ in. in diameter through their centres. These form guides for the legs of the bent wire N, and also avoid splashing of the oil.

It is tempting to use for the cups glass specimen tubes with ground stoppers, which can be inserted when the break is not in use, the screw R being taken out and the spring and wire removed to admit of their insertion. But, for the reason given above, it is on the whole decidedly better to replace the glasses by steel bicycle tubing. This has the advantage that, if desired, the current may be taken from the outside, and the bent wires dispensed with, and it also has the advantage that the light of the spark is hidden, which is very desirable in some experiments, besides the obvious advantage that no spark, however vigorous, can break the cup. A bottom can easily be made to the short lengths of steel tube by driving in little discs of iron with discs of cork above them to keep the mercury from escaping between the disc and the tube. Mercury and brass slowly amalgamate, but a long time elapses before this becomes serious. It must not be forgotten, however, that soft solder is dissolved by mercury in a few hours, and must on no account come in contact with it. It is impossible to prevent the formation of a thin mud, which almost immediately renders the insulating liquid turbid, but this does not impair the action of the break. It is well to have some means of adjusting the height of the cup in which the spark passes. This may either be done by a screw arrangement, or, more simply, by setting the cups stiffly in the rings which hold them, or, better, by the insertion of a wedge between the bottom of the cup and the baseboard. The cup into which the long wire dips may of course be of glass, as no spark plays in it.

XVI

HAND BREAKS

IT is often desirable—as in experiments in telegraphy by the Hertzian waves—to produce one spark only, and for this purpose to produce a sudden rupture of the primary circuit not followed by subsequent makes and breaks. In other words, we want a signalling key suitable to a coil. This is well effected by the hand break. A well-known form of this which was made by Mr. Apps for his large coils, and which was regularly used at the Polytechnic when the large coil was exhibited there, consists of a large vase or stoppered bottle containing mercury with alcohol over it, into which, when the stopper was removed, a vertical plunger tipped with platinum could be pressed down by the hand, and which on removal of the hand was suddenly forced up by a helical spring surrounding it. The instrument will be seen figured on page 45 of vol. ii. of Gordon's "Electricity and Magnetism," and doubtless in other electrical works. This, of course, answers all purposes, but the following design is better adapted to be clamped to the baseboard of a coil in which the system is adopted of making the breaks removable, as keeping the hand further removed from the danger of shocks than it

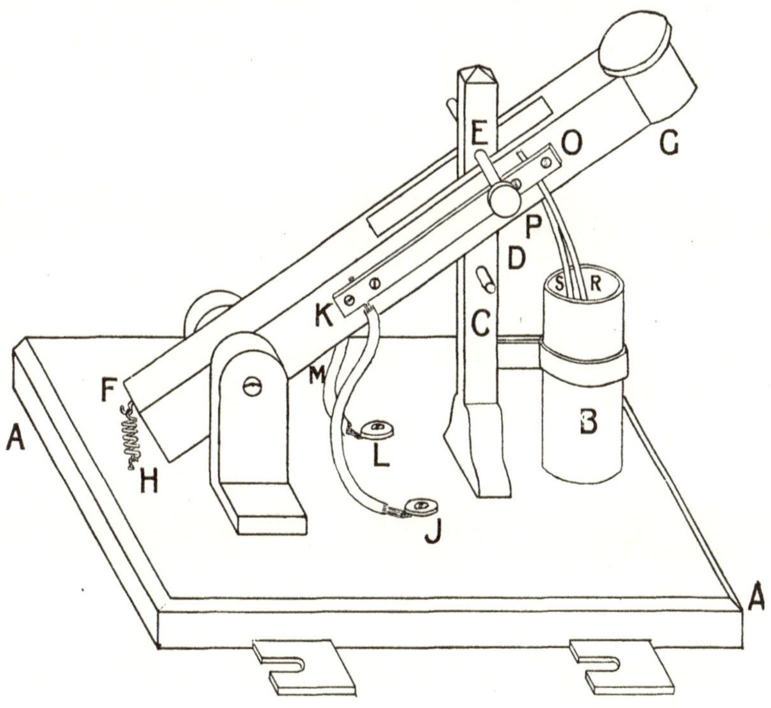

FIG. 31.—Hand Break.

would be if the vertical plunger were used. It is also exceedingly simple to make. In Fig. 31, A A is the baseboard with slotted brass strips as in the previously described breaks. B is a mercury cup made of a short piece of bicycle tubing closed at the bottom, or a gas-cap attached to the wooden post C. A wooden lever, F G, turns on a pin between two standards as shown, and its shorter end is pulled down by a helical spring F H. This lever has a slotted hole in it enabling it to pass down the post C, in which two stout pins are driven horizontally, the lower one D being fixed, and the upper one E being only hand-tight and being capable of being pulled out when required. It is well not to trust to the trunnions and standards to conduct the current, as the contacts would not be satisfactory, so the brass strips under the base are respectively connected by stout soldered connections with the two flexible leads, J K, L M, which again are soldered to two other strips of brass, only one of which, K O, is seen in the figure, attached one to each side of the lever, and these strips in turn are fastened by solder or clamping-screws to the two curved wires O R, P S, which should be of silver, or at least have short pieces of silver wire fastened to their ends by silver solder. When the break is not in use, the lever stands up with the curved wires clear of the mercury cup, and leaving room for a stopper to stand in its neck. But when it is required to use the break, it is clamped to the coil, and, the pin E having been for a moment removed, and the stopper taken out of the cup, the lever is pressed down and the pin E replaced. In this position the curved wires stand in the oil but clear of the mercury. To make contact, the end of the lever G (which may have an ivory finger-piece) is pressed down until the lever rests on the fixed pin D, and to break contact the finger is drawn off, when the lever flies up, pulling the curved wires out of the mercury, its motion being finally arrested by the pin E. It will be seen that there are in this break no bad contacts at all, as the current never has to pass from a journal to a bearing, and if the curved wires make good contact with the mercury the resistance of the apparatus is infinitesimal. Before leaving the subject of hand breaks, one remark may be made. It has been suggested that if it is desired to use a hand break in a coil where the spring break is not removable, the latter should be cut out by screwing up the contact screw. Now it is quite clear, that by screwing up the contact screw of the vibrating break the primary circuit is completed, and may be broken in any other place; but only the smallest effect can be produced with the coil in this condition, for by closing the vibrating break the condenser is short circuited and rendered useless. The only way to use a hand break with a coil of ordinary construction is not to close the spring break, but to open it, and attach leads to the two pillars of it, or to some fittings in metallic connection therewith respectively.

XVII

ELECTROLYTIC BREAKS

NO account of breaks to be used with an induction coil would be complete without some mention of the Electrolytic Interrupter recently devised by Dr. A. Wehnelt. Its action depends upon the discovery, said to have been made by Spottiswoode, that if a sufficently powerful electric current is passed through a cell containing a liquid electrolyte, the two electrodes being of very unequal area, the current is no longer continuous, but intermittent.

The application of this discovery to coils is as yet (June, 1899) too recent for one to feel certain that the best arrangement has been devised, but the following construction will answer. The condenser is disconnected from the coil, and the break removed or screwed up (the condenser being now unnecessary) in such a way that the two ends of the primary coil can have wires connected with them. Leads are taken from them to the two poles of a powerful battery or, through an appropriate resistance, to the street mains, if that source of current is available. In one of the leads an electrolytic cell is inserted consisting of a jar or beaker filled with dilute sulphuric acid (rather strong, for good conductivity), one electrode being a piece of sheet lead of any convenient size, say six or eight square inches in area, and the other being one

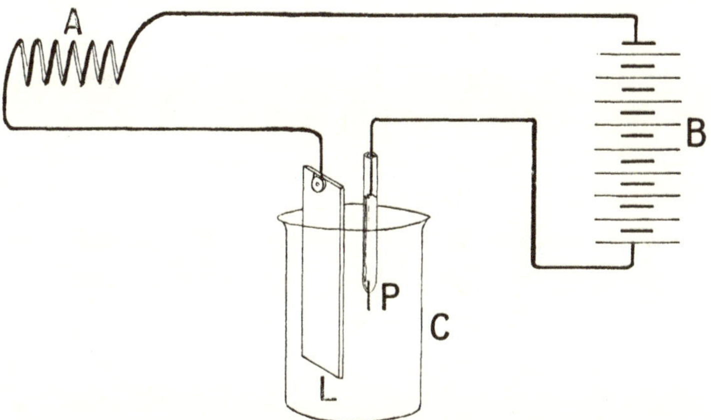

FIG. 32.—Diagram of Dr. Wehnelt's Electrolytic Break. A, primary of coil; B, battery; C, beaker of dilute sulphuric acid, in which are immersed L, leaden cathode, and P, platinum wire anode.

or more short platinum wires. The wires may be about No. 18 S.W.G., and they may project half an inch, more or less, into the liquid. The platinum wire or wires must be made the anode (see Fig. 32).

Things being so arranged, the current at once becomes intermittent, the period being very rapid and the interruption very complete. The number of interruptions per second remains very constant with a given arrangement of the apparatus, but varies with the voltage of the current, the size of the platinum wire-anode and the length to which this projects into the liquid. It may vary from about 1,000 to 3,000 breaks per second. It would be supposed that the effect produced in the secondary circuit would be insignificant, on account of the small time allowed for the magnetisation of the core, but experiment shows, on the contrary, that the effect is brilliant. Radiographs are taken with a short exposure and, according to D'Arsonval, the X-rays "have a penetrating power far beyond the usual." An alternating current may be used, but, nevertheless, the current in the primary is apparently unidirectional. For the production of ozone the Wehnelt interrupter seems very suitable, giving quantities of the gas "incomparably greater than the ordinary spring interrupter" (D'Arsonval).

One way of arranging the anode is to seal a short length of platinum wire by means of enamel into the end of a glass tube a few inches long, and to fix the tube in a clip-stand with the sealed end downwards and the projecting part of the wire dipping into the acid; then good contact can be secured by filling the glass tube with mercury and dipping the copper lead into it. Or the platinum wire may be passed tightly through a piece of ebonite used as a cork, which arrangement has the advantage that it allows the length of the wire which projects into the liquid to be varied.

The principal drawback to the use of this break is that a high voltage is necessary to work it. The actual minimum seems at present not to have been definitely ascertained, as by increasing the number of platinum points and warming the solution the necessary battery power can be reduced, but ordinarily a pressure of about 30 volts at least seems to be necessary, corresponding to about 15 secondary cells, or 17 cells of Grove.

The effect may be modified as to frequency, and often improved, by including more inductance in the circuit than the coil furnishes. If a thick coil of stout wire wound on a hollow bobbin be connected in series with the induction coil and break, and a loose iron core be inserted to a greater or less distance inside it, every change of position of the core modifies the frequency and changes the pitch of the note produced by the interrupters, forming a kind of electrical slide-trombone.

Mr. E. W. Caldwell has published in the *Electrical Review* of New York the discovery that this break may be modified in an interesting way. Instead of making one electrode much smaller than the other, he makes both of moderate size and both of lead, but divides the electrolytic bath by an insulating partition having one or more small holes in it. He points out that a simple way to effect this is to fuse a small hole at or near the bottom of a test-tube, in which is suspended, by a wire passing through a cork, one electrode. The tube is then allowed to stand in a beaker in which is placed the

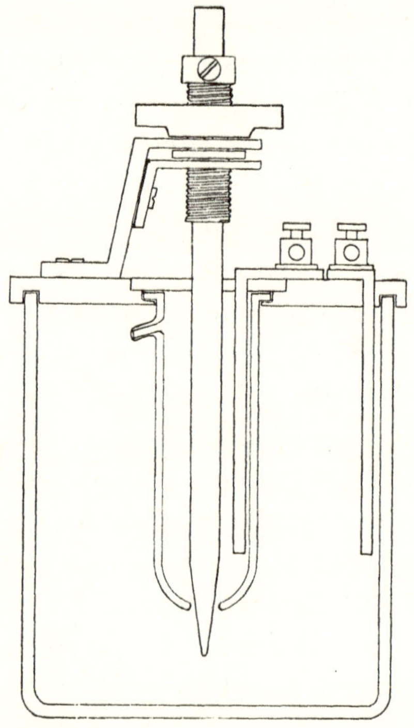

FIG. 33.—Swinton's Adjustable Caldwell Break.

other electrode. Beaker and tube are to be full of the liquid. He has tried various electrolytes, and prefers dilute sulphuric acid.

Mr. Swinton has arranged a convenient modification of the Caldwell break by which the resistance of the primary circuit and the speed of the interruption may be exactly regulated. It is shown in Fig. 33, and consists of an outer cylindrical vessel of glass in which is immersed one leaden electrode connected with a terminal brought out through an ebonite cover. From this ebonite cover hangs also an inner vessel like a large test-tube, but of thicker glass, containing the other leaden electrode. Communication is established between the two vessels by a hole fused in the bottom of the inner one. This hole is about ⅛ of an inch in diameter. So far the apparatus resembles Caldwell's, but the new feature is a vertical glass rod drawn out at the bottom to a long conical point like a cedar pencil. This rod passes centrally through the ebonite cover, and its conical point passes through the hole in the bottom of the inner vessel. By turning a milled head which surrounds the upper part of this glass rod and is threaded like a nut engaging a sleeve clamped to the glass rod, the latter is raised or lowered, causing its conical end to fill more or less the hole in the inner vessel, or, if desired, practically to close it altogether. By the use of this device the strength of the primary current and the rapidity of the interruptions can be exactly adjusted as desired. It will be noted that the Caldwell form, and equally Mr. Swinton's arrangement, being symmetrical with respect to the two electrodes, it is immaterial in which direction the current flows through them. The little spout seen near the top of the inner vessel is to provide an overflow for the acidulated water, which rises in it rapidly during use. The spout must be high enough to ensure that the overflow takes place by drops, and not by an unbroken stream, which would short-circuit the apparatus and prevent it from working.[1]

[1] See for further details: Wehnelt, *Electrotechn. Zeitschrift*, 20, pp. 76–78, 1899; D'Arsonval, *Comptes Rendus*, 128, pp. 529–532, 1899: *Science Abstracts*, April, 1899, pp. 241, 242; Carpentier, *Comptes Rendus*, April 17, 1899; Armagnat, *Comptes Rendus*, same number; *The Electrician*, May 5, 1899, p. 39, and June 2, 1899, p. 183; *The Electrical Review* (English), May 19, 1899, p. 837, and May 26, 1899, p. 874; Swinton, *Proc. Phys. Soc.*, June, 1899, p. 419, and *Electrician*, June 30, 1899, p. 332.

XVIII

CONSTANTS OF THE COIL

IT may be interesting in conclusion, and will enable the amateur coil-maker to judge of his work, to collect here some of the constants of a coil made in the way which has now been fully described.

CORE.

Wire gauge No. 22, S.W.G., length 18 in., diameter $1\frac{11}{16}$ in., weight 7 lbs. 13 ozs.

PRIMARY.

Gauge No. 14, S.W.G., single silk-covered H.C. copper.
First layer, 193 turns. Resistance, ·15 ohms (approx.).
Second layer, 189 turns. Resistance, ·15 ohms (approx.).
Third layer, 188 turns. Resistance, ·15 ohms (approx.).

RESISTANCES OF THE PRIMARY.

All in parallel, ·05 ohm.

Used as in two layers—out through one and back by the other two, ·225 ohm.

All in series, ·452 ohm.

Length of primary, 286 feet.

Weight, 5 lbs. 13 oz.

Main Ebonite Tube.

Length, 22 in.; thickness, $\frac{1}{4}$ in.; outside diameter, 3 in.

Condenser.

In two portions; sizes of foils, $15\frac{1}{2}$ in. by 9 in.
Number of foils, (*a*) 33, (*b*) 49; total 82 sheets.
Effective surface about 32 square feet.
Capacity about ·39 microfarad.
Insulation, two thicknesses of paraffined bank post paper.

Secondary.

Total weight of wire No. 36, 17 lbs.
Total weight of wire No. 34, 2 lbs. 1 oz.
Total number of turns, 79,168.
Total length about 17 miles.
Total resistance, 18,270 ohms.

In 96 sections, insulated by one thickness of blotting-paper paraffined, increased to two thicknesses where the P.D. is greatest. Outside diameter of secondary coil $6\frac{5}{8}$ in. at the middle, diminishing to $5\frac{1}{2}$ in. at the ends. Inside diameter $3\frac{1}{2}$ in. at the middle, increasing to 4 in. at ends.

Spark length, limited by position of terminals to $13\frac{1}{2}$ in.

APPENDIX I

On the Question, "How Many Layers of Primary Wire are Generally Desirable?"

As the question of the best number of layers of wire for the primary of an induction coil seems not to have been definitely settled, or at all events not to be generally known, it seemed worth while to make a few experiments on this point. But the matter is a complicated one, as so many other conditions may also be varied besides the one which it is desired to investigate, such as the voltage and total resistance of the battery employed, the size and build of the coil, the capacity of its condenser, and the use to which the coil is to be put. Want of time prevented all the combinations of these varying conditions being separately investigated, and therefore the experiments were made with one battery only, viz., six large Grove cells in good condition. The voltage of these may be taken to have been 11½, and the resistance not very different from one ohm.

The coil used was the one before referred to in this handbook, in which there are three layers of primary, the two ends of each of which are brought out to separate binding-screws. To these binding-screws was connected a set of six mercury cups formed by cavities in a block of wood, and other slips of wood were prepared having amalgamated copper wires fastened to them slightly projecting from the lower surface, and so arranged that by placing the appropriate slip upon the mercury cups, any of the following three arrangements could be produced in a moment.

First arrangement:—All three layers of primary in "parallel arc" so as to be equivalent to one layer only, of very low resistance. This may be called the "one" arrangement.

Second arrangement:—The current entering the primary by one terminal, passing through one layer, returning under the base, and then passing in parallel through the other two layers. (It will be remembered that all the layers are right-handed helices.) This is equivalent to working the coil with a primary of two layers of moderate resistance, and may be called the "two" arrangement.

Third arrangement:—The current passing in series through all the layers, as usual in coils. This may be called the "three" arrangement.

It then became necessary to select, from the numerous uses to which a coil may be put, some which should be typical, and should bring out strongly, if possible, whatever differences there might be in the behaviour of the coil under the varying conditions. It was decided to try four effects—(1) the maximum length of spark producible between a point and plate; (2) the heating effect of a short spark on a platinum wire attached to the negative electrode; (3) the excitation of a Röntgen tube, as tested on a fluorescent screen; and (4) the appearance in the spectroscope of the metallic lines of iron, aluminium, and magnesium.

In none of these uses was the third layer found to present any advantage. As regards length of spark, by proper management of the break, the coil would give the longest spark which the position of its secondary terminals permitted, with either two layers or three. With one layer the spark was barely an inch less.

With the short spark the heating effect on the platinum wire terminal was decidedly greater with the "one" arrangement than with either of the others, and of those the "two" arrangement was better than the "three."

With the Röntgen tube, and with the spectroscope, no advantage could be found in a third layer, and in any case such as the use of the spectroscope or the phosphorescent screen, where eye observations have to be made, the greater rapidity of action of the break when two layers instead of three are used, due to diminished inductance of the primary, is advantageous, as giving a more continuous impression on the eye.

The conclusion is that for this arrangement of battery for most purposes two layers are to be preferred, and for heating effects one. It must also not be forgotten that in a coil originally designed for two layers only the secondary can be brought nearer to the core by the space occupied by one whole layer and its insulation, while if the maker is content always to use two layers in series and never in parallel, some simplification results from the possibility of winding the inside as a right-handed helix, and the outer layer as a left-handed one, thus bringing out both ends of the primary at the same end of the coil and avoiding the necessity of making more than one ebonite pillar to convey these wires to the base. This pillar may be, at pleasure, either at the break end or the other end of the bobbin.

But if either for "calorific" or any other experiments it is desired to have the option of using the two layers as one, then the layers should both be wound in the same direction, *i.e.*, both as right- or both as left-handed helices, and the ends carried under the base and brought out at the end farthest from the break and attached to four terminals arranged in the manner previously recommended for six (see p. 110).

It will be noted that, in the experiments above described, as there were (approximately) 190 turns in each layer, the resistance of each layer being about ·150 ohm, that of the battery 1 ohm, and that of the leads negligible, the number of "ampère-turns" round the core in the three arrangements respectively were, for steady currents, 2,081, 3,567, and 4,520. The core must therefore have been twice as highly magnetised in arrangement "three" as in arrangement "one." With accumulators of low resistance instead of Grove cells, this difference would have been even more marked. Hence it is clear that the inferiority of three layers to two must be due to the increased self-induction in the "three" arrangement, and the consequent less rapid collapse of the magnetic lines on breaking circuit.

APPENDIX II

To Find the Capacity of a Condenser

There are several ways of doing this. Perhaps the most usual is with a "Ballistic Galvanometer." The method is fully described in the practical text-books (*e.g.*, Ayrton's "Practical Electricity," p. 328; Henderson's "Practical Electricity and Magnetism," vol. ii., p. 232). It consists in charging the condenser to a definite potential, discharging it through the galvanometer, noting the deflection, and afterwards finding what resistance has to be inserted in place of the condenser in order that the same difference of potential, or a known fraction of it, may produce a steady deflection equal to the throw of the needle previously noted. Full practical details of this method, with the formula for calculation, will be found in the works referred to, and elsewhere, and the method is capable of giving very accurate results when carefully applied.

But the following method, briefly described in Professor S. P. Thompson's "Elementary Lessons in Electricity and Magnetism," p. 425 (1895 edition), and probably elsewhere, is on the whole easier, and dispenses with the ballistic galvanometer, a tiresome piece of apparatus to work with.

The necessary apparatus comprises a tuning-fork maintained in vibration by an independent current, so arranged that a scrap of platinum attached to one prong vibrates between two other scraps mounted in such a manner as to impede the vibration of the fork as little as possible; also a Wheatstone Bridge and a sensitive galvanometer.

Fig. 34 shows the connections diagrammatically, V_1 V_2 V_3 V_4 represent the Wheatstone Bridge, the letters also standing for the potentials at the points figured.

B is the testing battery, G the galvanometer, C the condenser whose capacity is to be found, and F the tuning-fork vibrating between the platinum contacts, p p′.

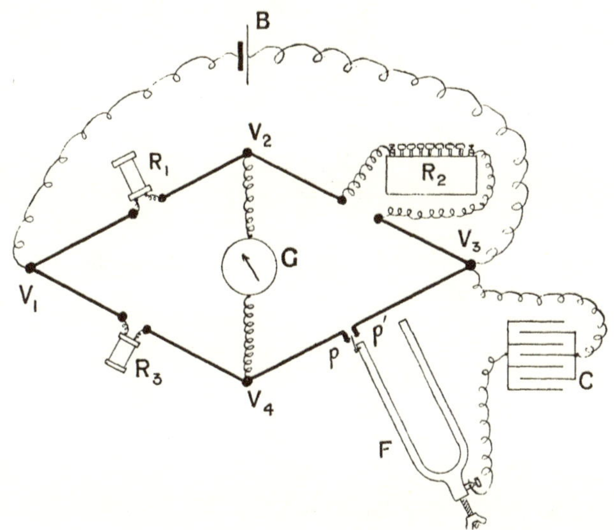

FIG. 34.—Diagram of Connections or Measuring Capacity of a Condenser. V_1 V_2 V_3 V_4, Wheatstone bridge; B, battery; C, condenser; G, galvanometer; R_1 R_3, ratio coils; R_2, resistance box; F, tuning-fork; p p′, platinum contacts. (Tapping-keys in the battery and galvanometer circuits not shown.)

R_1 and R_3 are known resistances; R_2 is the adjustable resistance, or box of coils.

The stem of the fork is in electrical connection with one side of the condenser, the other side of which is connected with the point V_3. It will be observed that each time the fork strikes against p the condenser is charged by the difference of potential between V_4 and V_3, and each time it strikes p′ the condenser is short-circuited and discharged. Consequently an intermittent stream of electricity is passing from V_4 to V_3, while a continuous one is passing from V_2 to V_3 through the resistance box. As the fork is vibrating rapidly, the needle of the galvanometer has no time to move between the separate impulses arising from the charging of the condenser, and it is possible to adjust the resistance R_2 until the needle is unaffected when the galvanometer key is depressed.

To obtain the formula, suppose for a moment that the condenser and fork are removed and replaced by a resistance R_4, and let C_1 C_2 C_3 C_4 be the currents through R_1 R_2 R_3 R_4 respectively. Then by Ohm's law—

$$\begin{aligned} V_2 - V_1 &= C_1 R_1 \\ V_3 - V_2 &= C_2 R_2 \\ V_3 - V_4 &= C_4 R_4 \\ V_4 - V_1 &= C_3 R_3 \end{aligned}$$

But since R_2 has been so adjusted that the galvanometer is unmoved, we must have

$$V_2 = V_4$$

$$\text{so that } \left.\begin{aligned} V_2 - V_1 &= C_1 R_1 = C_3 R_3 \\ V_3 - V_2 &= C_2 R_2 = C_4 R_4 \end{aligned}\right\} \quad . \; . \; . \; . \qquad (1)$$

$$\text{and } C_1 = C_2,\ C_3 = C_4. \; . \; . \; . \qquad (2)$$

$$\text{Therefore } R_4 = \frac{R_3}{R_1} R_2 \; . \; . \; . \; . \qquad (3)$$

Now suppose R_4 replaced by the condenser whose capacity we want to measure, attached to a fork making N complete vibrations per second. Then the formula for a condenser being

$$K = \frac{A}{V},$$

where K represents the capacity, A the charge, and V the difference of potential of its two sides, the number of coulombs passing per second, *i.e.*, the effective current in the branch $V_4 V_3$ is

$$C_4 = N A$$
$$= N K (V_3 - V_4)$$

$$\text{whence } V_3 - V_4 = \frac{C_4}{N K}$$

$$\text{But from (1) } V_3 - V_2 = C_4 R_4,$$
$$= V_3 - V_4.$$

$$\text{Hence, by division } R_4 = \frac{1}{N K}.$$

$$\text{Therefore from (3) } \frac{1}{N K} = \frac{R_3}{R_1} R_2$$

$$\text{or } K = \frac{R_1}{R_3} \frac{1}{N R_2}.$$

If the condenser is not too small, R_1 may be made equal to R_3, and we get the very simple formula

$$K = \frac{1}{N R_2}.$$

As we are measuring the currents in amperes and the resistances in ohms, the result comes out in farads, and must be multiplied by a million, giving us finally—

$$\text{Capacity in microfarads} = \frac{1{,}000{,}000}{N R_2}.$$

A simple way of arranging the contacts in such a way as not to disturb the period of the fork is to cut a narrow strip of thin sheet ebonite, and attach to one end of it, on opposite sides, by little screws or rivets, little pieces of thin platinum wire projecting about an eighth of an inch beyond the end. The heads of the little screws also hold down short lengths of No. 36 copper wire leading to a couple of binding-screws at the other end of the strip. A cork or a cylindrical piece of wood attached by nails to the ebonite serves to hold the apparatus in a clip-stand, and allow of its height being adjusted to suit the fork. If the ebonite is $\frac{1}{16}$ in. thick, $\frac{3}{8}$ in. wide, and projects from its support four or five inches, it will oppose so slight an impediment to the motion of the fork—if the latter is fairly massive—that it may be assumed that its period is unchanged.

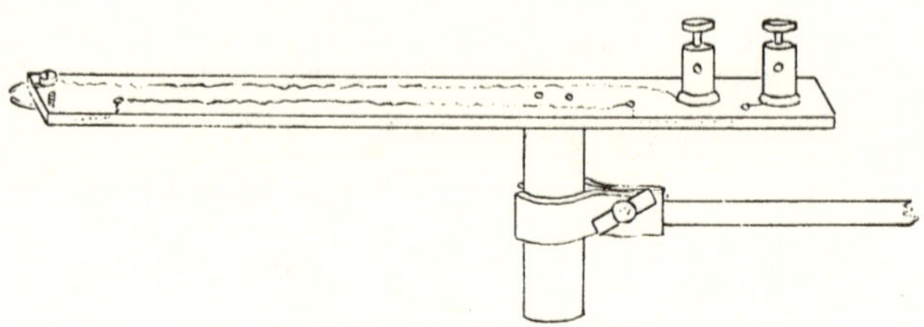

FIG. 35.—Slip of Ebonite, with scraps of platinum wire projecting rom the end, forming the contacts p p′ in figure 34.

It should be noted that if the fork is of French make the figures stamped upon it represent the number of *single* vibrations per second, and this number must be halved to give N.

Fig. 35 shows the contact-apparatus set up.

INDEX

UNWIN BROTHERS, THE GRESHAM PRESS, WOKING AND LONDON.

www.ingramcontent.com/pod-product-compliance
Lightning Source LLC
LaVergne TN
LVHW091641100826
845152LV00006B/126/J